Memoirs of the American Mathematical Society
Number 368

Michael B. Marcus

ξ-radial processes and random Fourier series

Published by the
AMERICAN MATHEMATICAL SOCIETY
Providence, Rhode Island, USA

July 1987 · Volume 68 · Number 368 (first of 3 numbers)

MEMOIRS of the American Mathematical Society

SUBMISSION. This journal is designed particularly for long research papers (and groups of cognate papers) in pure and applied mathematics. The papers, in general, are longer than those in the TRANSACTIONS of the American Mathematical Society, with which it shares an editorial committee. Mathematical papers intended for publication in the Memoirs should be addressed to one of the editors:

Ordinary differential equations, partial differential equations, and applied mathematics to JOEL A. SMOLLER, Department of Mathematics, University of Michigan, Ann Arbor, MI 48109

Complex and harmonic analysis to ROBERT J. ZIMMER, Department of Mathematics, University of Chicago, Chicago, IL 60637

Abstract analysis to VAUGHAN F. R. JONES, September 1986–July 1987: Institut des Hautes Études Scientifiques, Bures-Sur-Yvette, France 91440

Classical analysis to PETER W. JONES, Department of Mathematics, Box 2155 Yale Station, Yale University, New Haven, CT 06520

Algebra, algebraic geometry, and number theory to DAVID J. SALTMAN, Department of Mathematics, University of Texas at Austin, Austin, TX 78713

Geometric topology and general topology to JAMES W. CANNON, Department of Mathematics, Brigham Young University, Provo, UT 84602

Algebraic topology and differential topology to RALPH COHEN, Department of Mathematics, Stanford University, Stanford, CA 94305

Global analysis and differential geometry to JERRY L. KAZDAN, Department of Mathematics, University of Pennsylvania, E1, Philadelphia, PA 19104-6395

Probability and statistics to RONALD K. GETOOR, Department of Mathematics, University of California at San Diego, La Jolla, CA 92093

Combinatorics and number theory to RONALD L. GRAHAM, Mathematical Sciences Research Center, AT&T Bell Laboratories, 600 Mountain Avenue, Murray Hill, NJ 07974

Logic, set theory, and general topology to KENNETH KUNEN, Department of Mathematics, University of Wisconsin, Madison, WI 53706

All other communications to the editors should be addressed to the Managing Editor, LANCE W. SMALL, Department of Mathematics, University of California at San Diego, La Jolla, CA 92093.

PREPARATION OF COPY. Memoirs are printed by photo-offset from camera-ready copy prepared by the authors. Prospective authors are encouraged to request a booklet giving detailed instructions regarding reproduction copy. Write to Editorial Office, American Mathematical Society, Box 6248, Providence, RI 02940. For general instructions, see last page of Memoir.

SUBSCRIPTION INFORMATION. The 1987 subscription begins with Number 358 and consists of six mailings, each containing one or more numbers. Subscription prices for 1987 are $227 list, $182 institutional member. A late charge of 10% of the subscription price will be imposed on orders received from nonmembers after January 1 of the subscription year. Subscribers outside the United States and India must pay a postage surcharge of $25; subscribers in India must pay a postage surcharge of $43. Each number may be ordered separately; *please specify number* when ordering an individual number. For prices and titles of recently released numbers, see the New Publications sections of the NOTICES of the American Mathematical Society.

BACK NUMBER INFORMATION. For back issues see the AMS Catalogue of Publications.

Subscriptions and orders for publications of the American Mathematical Society should be addressed to American Mathematical Society, Box 1571, Annex Station, Providence, RI 02901-9930. *All orders must be accompanied by payment.* Other correspondence should be addressed to Box 6248, Providence, RI 02940.

MEMOIRS of the American Mathematical Society (ISSN 0065-9266) is published bimonthly (each volume consisting usually of more than one number) by the American Mathematical Society at 201 Charles Street, Providence, Rhode Island 02904. Second Class postage paid at Providence, Rhode Island 02940. Postmaster: Send address changes to Memoirs of the American Mathematical Society, American Mathematical Society, Box 6248, Providence, RI 02940.

The paper used in this journal is acid-free and falls within the guidelines established to ensure permanence and durability. ∞

TABLE OF CONTENTS

Preface   1

1. Introduction   4

2. Representing ξ-radial processes 26

3. Necessary conditions for continuity 46

4. Sufficient conditions for continuity 60

5. Processes for which the Levy transforms or the logarithms of
    the characteristic functions are regularly varying with index
    $1 < p < 2$ 94

6. Processes for which the Levy transforms or the logarithms of
    the characteristic functions are regularly varying with index
    1 or 2 131

7. Suprema of ξ-radial processes and random Fourier series . . . 163

ABSTRACT: A ξ-radial process is a stochastic process whose finite joint distributions are defined in terms of a symmetric real valued infinitely divisible random variable ξ. Let $E \exp i\lambda\xi = \exp -\Psi(|\lambda|)$, $-\infty < \lambda < \infty$. Let G be a real Abelian compact group, Γ be the dual group of G, $\gamma \in \Gamma$, m a probability measure on Γ and E_m expectation with respect to m. A real valued stochastic process $\{X(t)\}_{t\in G}$ is called ξ-radial if for all $\alpha_1, \ldots, \alpha_n \in \mathbb{R}$, $t_1, \ldots, t_n \in G$

$$(0.1) \qquad E \exp i \sum_{j=1}^{n} \alpha_j X(t_j) = \exp\left(-E_m \Psi\left(\left| \sum_{j=1}^{n} \alpha_j \gamma(t_j) \right|\right)\right) .$$

Integral conditions are obtained both for the continuity and unboundedness a.s. of $\{X(t)\}_{t\in G}$ in terms of the metric entropy of G with respect to various metrics or pseudo-metrics related to Ψ and m. These results extend those of the author and G. Pisier for strongly stationary p-stable processes, i.e. when $\Psi(|\lambda|) = |\lambda|^p$, $0 < p < 2$.

The same methods used to study strongly stationary ξ-radial processes give results about the more classical random Fourier series

$$Y(t) = \sum_{\gamma \in A} a_\gamma \xi_\gamma \gamma(t), \quad t \in G$$

where $A \subset \Gamma$ is countable, $a_\gamma \in \mathbb{C}$ and $\{\xi_\gamma\}_{\gamma\in A}$ are independent identically distributed random variables with characteristic function $E \exp i\lambda\xi_\gamma = \psi(|\lambda|)$, $-\infty < \lambda < \infty$. If $\psi(|\lambda|) \sim \phi(|\lambda|)$ as $\lambda \to 0$ where $\phi(|\lambda|)$ is convex with $\phi(0) = 0$, integral conditions for the continuity of $\{Y(t)\}_{t\in G}$ are given in terms of the metric or pseudo-metric

$$d_{Y,\phi}(s,t) = \inf\{c: \sum_{\gamma \in A} \phi\left(\frac{|a_\gamma(Y(s) - Y(t))|}{c} \right) \le 1\} .$$

When $\phi(|\lambda|) = \psi(|\lambda|) = |\lambda|^p$, $0 < p < 2$, the processes defined in (0.2) are special cases of those defined in (0.1), i.e. those for which m has discrete support. However, in general, the processes defined by (0.1) and (0.2) are not the same.

1980 Mathematics Subject Classification:

60G10, 60G17, 60G15, 60E07, 42A61, 42A20, 43A50

Library of Congress Cataloging-in-Publication Data

Marcus, Michael B.
    [xē]-radial processes and random Fourier series.

(Memoirs of the American Mathematical Society, ISSN 0065-9266;
no. 368)
    On t.p. "[xē]" appears as the Greek character.
    Bibliography.
    1. Stationary processes. 2. Fourier analysis. 3. Sampling (Statistics)
I. Title. II. Series.
QA3.A57   no. 368   510 s   [519.2′32]   87-12569
[QA274.3]
ISBN 0-8218-2432-5

for Jane

PREFACE

This monograph is a study of the sample path continuity of a certain class of stationary stochastic processes. It continues a line of research that was initiated by the beautiful Theorem of Dudley and Fernique, in which a necessary and sufficient condition for the continuity of stationary Gaussian processes is given in terms of an integral condition involving the metric entropy of the L^2 metric of the increments of the process. This result was a breakthrough in the study of stochastic processes because it shifted attention from conventional metrics on the domain of the process, such as Euclidean distance in R^n, to a metric more intrinsically associated with the process itself. Many questions in the study of path properties of stochastic processes were reexamined in the light of metric entropy and several important results were obtained. One particularly satisfying result of this nature is that the same entropy condition of Dudley and Fernique is also a necessary and sufficient condition for the uniform convergence a.s. of Rademacher random Fourier series. This is due to Gilles Pisier and the author. It suggested that metric entropy conditions and the theory of Gaussian processes could be useful in harmonic analysis as was subsequently shown in many celebrated papers of Pisier.

The results on random Fourier series in the monograph Random Fourier series with applications to harmonic analysis by G. Pisier and the author are valid, generally speaking, whenever the independent random coefficients have finite second moments. It is natural to wonder what can be said when the coefficients of the series do not have second moments. An investigation in this direction was undertaken by J. Cuzick and T. L. Lai. In a later paper the author and G. Pisier showed that an integral condition involving the metric entropy of a certain L^p metric gives necessary and sufficient conditions for continuity of random Fourier series with p-stable coefficients when $1 < p < 2$. This result is also a natural extension of the Dudley-Fernique theorem for Gaussian processes which corresponds to the case $p = 2$.

An important component of the work on p-stable processes is a representation of these processes due to R. LePage. This representation allows one to consider a large class of stationary p-stable processes, not only random

Received by the editor May 10, 1985. Revised manuscript completed September 15, 1986. This research was supported in part by a grant from the National Science Foundation.

1

Fourier series with p-stable coefficients. Moreover LePage's representation
applies to all infinitely divisible processes. Another and different way to
generalize the classical random Fourier series is to let them have random
infinitely divisible coefficients. Both the author and G. Pisier realized
that the methods developed to study stable processes could also be used to
study both of these more general classes of processes. Some preliminary
joint investigations initiated the research for this manuscript.

The overall approach of this monograph closely follows the paper
"Characterizations of almost surely continuous p-stable random Fourier
series and strongly stationary processes" by the author and G. Pisier. Many
technical difficulties arise because of the greater generality of this
study. Nevertheless several sufficent and necessary conditions for continuity are obtained. In many special cases these conditions are quite close
but in no case, other than that of stable processes for which the results
were already obtained in the paper with Pisier, do we obtain a single
necessary and sufficient condition for continuity. This has led us to look
at many examples so that we could evaluate our results. The examples show
that some of our results are "best possible". They also lead us to a
conjecture about what may be a necessary and sufficient condition for the
continuity a.s. of ξ-radial processes which is given at the end of Chapter
1. This conjecture is not stated with great conviction. It is meant rather
to focus energy for further research. However, the examples show that if
the conjecture is false, then there is no metric entropy integral condition
for the continuity of ξ-radial processes except when they are stable. If
such a condition does not exist it would not be too surprising. We show in
this monograph that there is no such condition for random Fourier series
formed with an independent identically distributed sequence of infinitely
divisible random variables, except when the random variables are stable.

The reader knowledgeable in stochastic processes is no doubt aware of
the recent extraordinary work of M. Talagrand. We do not discuss
majorizing measures here because, since our metrics are translation invariant
on locally compact groups, the majorizing measure can be taken to be Haar
measure and continuity conditions can be expressed in terms of metric
entropy.

Those of us who study path properties of broad classes of stochastic
processes are sometimes asked if such studies have any use beyond their
aesthetic appeal. ξ-radial processes and random Fourier series can be
represented as Fourier transforms of independent increment processes.
When the independent increment process is time changed Brownian motion,
the Fourier transform is related to the problem of passing "white noise"

through a linear filter. There is a well developed theory concerning this which is used by engineers all the time. Whether or not it makes sense to consider Fourier transforms of other kinds of "noise" based on different independent increment processes, I don't know. The only way to find out is to discover what are the path properties of these new processes and to see how they compare with empirical observations. Therefore, from a practical point of view, a monograph such as this one can be looked upon as an initial step in the development of new models for physical phenomena.

I am deeply indebted to Susan Trussell and Pam Paholek for their excellent work in preparing this manuscipt.

1. INTRODUCTION

In [MP2] necessary and sufficient conditions are given for the continuity of the sample paths of random Fourier series with independent p-stable coefficients and for random Fourier transforms with respect to random independently scattered p-stable measures, $1 < p < 2$, (when $p < 1$ the results are trivial). When $p = 2$ these stochastic processes are Gaussian; the question of the continuity of their sample paths is resolved by the results of Dudley [D] and Fernique [F1], (see also [JM]). In this monograph the results of [MP2] are extended in two directions. We consider certain infinitely divisible measures on the space of continuous functions that are more general than p-stable measures and we consider random Fourier series with independent coefficients that are more general than p-stable random variables. These two classes of processes are not the same yet the same proofs suffice to give results in both cases. We now state the main results of this monograph beginning with those for infinitely divisible measures on the space of continuous functions which are also random Fourier transforms.

Let G be a locally compact Abelian group with dual group Γ and let K be a fixed compact neighborhood of the unit element of G. Let m be a finite positive Radon measure on Γ satisfying $m(\Gamma) = 1$ and let E_m represent the expectation operator on the probability space (Γ, G, m) where G is the Borel field generated by the continuous functions $\{\gamma(t); t \in K, \gamma \in \Gamma\}$.

We will study a class of infinitely divisible processes based on a symmetric real valued infinitely divisible random variable ξ with characteristic function

(1.1)
$$Ee^{iu\xi} = e^{-\Psi(|u|)}, \qquad -\infty < u < \infty,$$

4

where

(1.2) $$\Psi(|u|) = \int_0^\infty (\cos tu - 1) d\tau[\tau, \infty)$$

for τ a Levy measure on $(0, \infty)$. These processes will be called ξ-radial

processes; motivation for this nomenclature will be given in Section 2.

To avoid confusion we will define these processes first when Γ consists

only of real valued functions and then when Γ can contain complex valued

functions.

Case 1: Γ consists only of real valued functions. Let $\mathbb{R}^K$ denote the

real numbers indexed by K. We consider the class of infinitely divisible

measures on $\mathbb{R}^K$, or equivalently, of real valued stochastic processes

$\{X(t)\}_{t \in K}$ with characteristic functionals given by

(1.3) $$E \exp i \sum_{j=1}^n \alpha_j X(t_j) = \exp\left(-E_m \Psi\left(\left|\sum_{j=1}^n \alpha_j \gamma(t_j)\right|\right)\right)$$

for all $\alpha_1, \ldots, \alpha_n \in \mathbb{R}$, $t_1, \ldots, t_n \in K$ and all n, where Ψ is as

given in (1.1) and (1.2). Such a process will be called a real valued

strongly stationary ξ-radial process. Precise definitions will be given

in Section 2.

The processes $\{X(t)\}_{t \in K}$ can be expressed as random Fourier

transforms with respect to an independently scattered infinitely divisible

random measure M on (Γ, G, m). M is defined as follows: For

measurable sets $A \subset \Gamma$, $B \subset \Gamma$

$$E \exp i\lambda M(A) = \exp - m(A)\Psi(|\lambda|), \quad -\infty < \lambda < \infty$$

and, if $A \cap B = \emptyset$, $M(A)$ and $M(B)$ are independent. The stochastic

process

(1.4) $$\int_\Gamma \gamma(t) M(d\gamma), \quad t \in K$$

has the same characteristic functional as $\{X(t)\}_{t \in K}$. Furthermore,

Rosinski [R] has recently shown that if (1.3) is the characteristic functional of a probability measure on $C(K)$, (the Banach space of real valued continuous functions on K equipped with the sup-norm) or equivalently if $\{X(t)\}_{t\epsilon K}$ has a version with continuous sample paths then (1.4) can be defined in the usual way as a $C(K)$ valued stochastic integral.

Case 2: Γ contains complex valued functions. Let $\mathbb{C}^K$ denote the complex numbers indexed by K. We consider the class of infinitely divisible measures on $\mathbb{C}^K$, or equivalently, of complex valued stochastic processes $\{X(t)\}_{t\epsilon K}$ with characteristic functionals given by

$$(1.5) \qquad E \exp i \ \mathrm{Re}\Big(\sum_{j=1}^{n} \overline{\alpha}_j X(t_j)\Big) = \exp\Big(-E_m \ \Psi\Big(\Big| \sum_{j=1}^{n} \overline{\alpha}_j \gamma(t_j)\Big|\Big)\Big)$$

for all $\alpha_1, \ldots, \alpha_n \ \epsilon \ \mathbb{C}$, $t_1, \ldots, t_n \ \epsilon \ K$ and all n, where Ψ is as given in (1.1) and (1.2). Such a process will be called a complex valued strongly stationary ξ-radial process. It is clear that not all functions Ψ, related to symmetric infinitely divisible random variables ξ as in (1.1), allow a representation as in (1.5). A description of those that do allow such a representation is given in Remark 2.5.

In this case as well $\{X(t)\}_{t\epsilon K}$ can be expressed as a random Fourier transform with respect to a complex valued infinitely divisible independently scattered random measure on $(\Gamma, \ G, \ m)$. This will be discussed in Remark 2.5.

Let $\Phi: \mathbb{R}^+ \to \mathbb{R}^+$, $\Phi(0) = 0$, be convex. For such a function Φ we associate with $\{X(t), \ t \ \epsilon \ K\}$ as defined in (1.3) or (1.5) the Orlicz pseudo-norm

$$(1.6) \qquad d_{X,\Phi}(s,t) = \inf\Big\{c > 0: \ E_m \Phi\Big(\frac{|\gamma(s) - \gamma(t)|}{c}\Big) \leq 1\Big\} \ ,$$

for all $s, \ t \ \epsilon \ K$. When $\Phi(x) = x^p$ we will denote $d_{X,\Phi}$ by $d_{X,p}$.

Let $N(K,d;\epsilon)$ denote the smallest number of open balls of radius ϵ,

in the pseudo-metric d which cover K. (We will always assume that K
is metrizable.) For a non-decreasing function $H: \mathbb{R}^+ \to \mathbb{R}^+$ with
$H(0) = 0$, we define

(1.7)
$$J(H,d) = J(H,d,K)$$
$$= \int_0^\infty H\big(\log N(K,d;\epsilon)\big)d\epsilon \; .$$

The functions H that will interest us in Theorem 1.1 will be
defined in terms of an auxiliary function η. Let $\eta(x)$, $x \geq 1$, be a
continuous strictly increasing function such that

(1.8)
$$\int_1^\infty \frac{ds}{\eta^{-1}(s)} = \infty \; .$$

(This necessarily implies $\lim_{x\to\infty} \eta(x) = \infty$.) For such a function η we
define a function H_η as follows

(1.9)
$$H_\eta(x) = \begin{cases} \int_1^x \dfrac{ds}{\eta^{-1}(s)} \; , & x \geq 1 \; , \\ \\ 0 & , \quad 0 \leq x < 1 \end{cases} \; .$$

The next theorem pertains to the continuity of the processes defined
in (1.3) and (1.5).

Theorem 1.1: Let $\{X(t)\}_{t\in K}$ be an $\mathbb{R}^K$ ($\mathbb{C}^K$) valued stochastic process
with characteristic functional given by (1.3) ((1.5)) and let
$\eta: [1,\infty) \to \mathbb{R}^+$, $\eta(1) > 0$, be a strictly increasing function.
I.) Suppose there exist constants k_1', $k_2' > 0$ and real numbers x_0', x_1'
and y_1' and a real valued function $T(x)$ such that

(1.10)
$$\Psi(|x|) \geq k_1' T(x), \quad \forall x \geq x_0' \geq 0$$

and

(1.11)
$$T(xy) \geq k_2' \eta(x)\Phi(y), \quad \forall y \geq y_1' \geq 1, \; x \geq x_1' \geq 1$$

where Φ is as given above. Then

(1.12) $J(H_\eta, d_{X,\Phi}) = \infty$

implies that $\sup_{t \in K} |X(t)|$ is unbounded a.s.

II.) Suppose that in addition to the information given prior to Part I.)
$\eta(x)$ is regularly varying at infinity with index p, $1 < p < 2$, and
satisfies

(1.13) $\eta^{-1}(x)$ and $\dfrac{x}{\eta^{-1}(x)}$ are concave for $x \geq 1$.

Suppose also that $\Psi(|x|)$ is regularly varying at infinity with index
$1 < p < 2$ and that there exist constants k_1, $k_2 < \infty$ and a real number
$x_0' \geq 0$ and a strictly increasing real valued function $T(x)$ such that

(1.14) $\Psi(|x|) \leq k_1 T(x)$, $\forall x \geq x_0' \geq 0$

where

(1.15) $T(xy) \leq k_2 \eta(x)\Phi(y)$, $\forall y \geq 1$, $x \geq 1$.

Then, if

(1.16) $J(H_\eta, d_{X,\Phi}) < \infty$

$\{X(t)\}_{t \in K}$ has a version with continuous sample paths a.s.

The second class of processes which we will consider is the more
classical random Fourier series with independent coefficients. Let ξ be
a symmetric real valued random variable with characteristic function

(1.17) $E\, e^{i\lambda\xi} = e^{-\psi(|\lambda|)}$.

Let $A \subset \Gamma$ be countable and let $\{\xi_\gamma\}_{\gamma \in A}$ be independent copies of ξ
indexed by the elements $\gamma \in A$. Similarly let $\{a_\gamma\}_{\gamma \in A}$ be complex
numbers such that $\sum_{\gamma \in A} a_\gamma \xi_\gamma$ converges. We define

(1.18) $Y(t) = \sum_{\gamma \in A} a_\gamma \xi_\gamma \gamma(t)$, $t \in K$

as a $\mathbb{C}^K$ valued stochastic process. (If $\sum_{\gamma \in A} a_\gamma \xi_\gamma$ does not converge then (1.13) can not even be defined let alone have a version with continuous paths.)

Let $\phi: R^+ \to R^+$, $\phi(0) = 0$, be a strictly increasing convex function that is regularly varying at 0 with index p, $1 \le p \le 2$, and that satisfies

$$(1.19) \qquad \lim_{x \to 0} \frac{\phi(x)}{x^2} \ge c_1 \quad \text{and} \quad \overline{\lim_{x \to \infty}} \frac{\phi(x)}{x^2} \le c_2 \;,$$

for $c_1 > 0$, $c_2 < \infty$. Furthermore assume that $\{a_\gamma\}_{\gamma \in A}$ is such that

$$(1.20) \qquad \sum_{\gamma \in A} \phi(|a_\gamma|) < \infty \;.$$

We associate with $\{Y(t)\}_{t \in K}$ the Orlicz sequence space pseudo-norm

$$(1.21) \qquad d_{Y,\phi}(s,t) = \inf\left\{c > 0: \sum_{\gamma \in A} \phi\left(\frac{|a_\gamma||Y(s) - Y(t)|}{c}\right) \le 1\right\}$$

for all $s, t \in K$. When $\phi(x) = x^p$ we will denote $d_{Y,\phi}$ by $d_{Y,p}$. Let $\nu(x)$, $0 \le x \le 1$ be a continuous strictly increasing function such that

$$(1.22) \qquad \int_1^\infty \nu^{-1}\left(\frac{1}{s}\right) ds = \infty \;.$$

(This necessarily implies that $\lim_{x \to 0} \nu(x) = 0$.) For such a funcion ν we define a function $\overline{H}_\nu$ as follows

$$(1.23) \qquad \overline{H}_\nu(x) = \begin{cases} \int_1^x \nu^{-1}\left(\frac{1}{s}\right) ds & , \quad x \ge 1 \\ \\ 0 & , \quad 0 \le x \le 1 \end{cases} \;.$$

The next theorem pertains to the continuity of the processes defined in (1.17) and (1.18).

Theorem 1.2: Let $\{Y(t)\}_{t \varepsilon K}$ be as given in (1.18) and let $\nu(x)$,
$0 \leq x \leq 1$, $\nu(0) = 0$, be a strictly increasing function.

I.) Suppose there exist constants k_1', $k_2' > 0$ and some real number
$x_0 > 0$ such that

(1.24) $\psi(|x|) \geq k_1'\phi(x)$, $0 \leq x \leq x_0$

and

(1.25) $\phi(xy) \leq k_2'\nu(x)\phi(y)$, $y \geq 0$, $0 \leq x \leq 1$.

Then if

(1.26) $J(\overline{H}_\nu,\, d_{Y,\phi}) = \infty$,

$\sup_{t \varepsilon K} |Y(t)|$ is unbounded a.s. (See (1.7), (1.21) and (1.23) for the
definition of the term in (1.26)).

II.) Suppose that in addition to the information given prior to Part I.)
$\nu(x)$ is normalized so that $\nu(1) = 1$ and is regularly varying at zero
with index p, $1 < p < 2$, and satisfies

(1.27) $\dfrac{1}{\nu^{-1}\left(\frac{1}{x}\right)}$ is concave, $x \geq 1$

and

(1.28) $x\nu^{-1}\left(\frac{1}{x}\right)$ is concave, $x \geq 1$.

Suppose also that $\psi(|x|)$ is regularly varying at zero with index
$1 < p < 2$ and that there exist constants $0 < k_1$, $k_2 < \infty$ and some real
number $x_0 > 0$ such that

(1.29) $\psi(|x|) \leq k_1\phi(x)$, $0 \leq x \leq x_0$,

and

(1.30) $\phi(xy) \geq k_2\nu(x)\phi(y)$, $y \geq 0$, $0 \leq x \leq 1$.

(By (1.29) and (1.20) we see that $\sum_{\gamma \in A} a_\gamma \xi_\gamma$ converges thus $\{Y(t)\}_{t\epsilon K}$ is well defined.)   Then if

$$(1.31) \qquad\qquad J(\overline{H}_\nu,\, d_{\gamma,\phi}) < \infty$$

the series representing $\{Y(t)\}_{t\epsilon K}$ given in (1.18) converges uniformly a.s. and is a continuous version of $\{Y(t)\}_{t\epsilon K}$

To help explain Theorem 1.1 let us suppose that $T(x)$ in I.) and II.), besides being strictly increasing, satisfies $T(1) = 1$ and is regularly varying at infinity with index $1 < p < 2$. Thus we can write

$$(1.32) \qquad\qquad T(x) = x^p\, L(x)$$

where L satisfies $\lim_{y\to\infty} \dfrac{L(xy)}{L(y)} = 1$, $\forall x > 0$, i.e. L is a slowly varying function at infinity. Let us consider various conditions which T might satisfy.

$$(1.33) \qquad\qquad T(xy) \geq C\, x^p\, T(y), \qquad \forall x,y \geq 1$$

$$(1.34) \qquad\qquad T(xy) \geq C\, T(x)\, T(y), \qquad \forall x,y \geq 1$$

$$(1.35) \qquad\qquad T(xy) \leq c\, x^p\, T(y), \qquad \forall x,y \geq 1$$

$$(1.36) \qquad\qquad T(xy) \leq c\, T(x)\, T(y), \qquad \forall x,y \geq 1$$

Here C and c are constants not necessarily the same in each appearance. Note that if $L(x)$ is increasing (decreasing) in (1.32) then (1.33) ((1.35)) holds. On the other hand if $T(x) = (ax)^p (\log (ax))^\beta$ then (1.36) ((1.34)) holds if $\beta \geq 0$, ($\beta \leq 0$) if a is taken to be sufficiently large. (One might also note that if $T(x) = x^p$ then all the inequalities (1.33)–(1.36) can be taken to be equalities with $C = c = 1$.).

The next Corollary is obtained from Theorem 1.1 by taking $\eta(x) = \Phi(x) = T(x)$ (in (1.11) and (1.15)) or $\eta(x) = T(x)$ and $\Phi(x) = x^p$ or $\eta(x) = x^p$ and $\Phi(x) = T(x)$, in an obvious way, depending on which of the conditions (1.33) to (1.36) is satisfied. (In all that follows we write $a \sim b$ if there exist constants $0 < c_1 < c_2 < \infty$ such that $c_1 a < b < c_2 a$.)

<u>Corollary 1.3</u>: Let $\{X(t)\}_{t\in K}$ be an $\mathbb{R}^k(\mathbb{C}^k)$ valued stochastic process with characteristic functional given by (1.3) ((1.5)). Assume that $\Psi(x) \sim T(x)$ for x sufficiently large where $T(x)$ is a convex function, $T(0) = 0$, which is regularly varying at infinity with index $1 < p < 2$ and let $\frac{1}{p} + \frac{1}{q} = 1$.

I.) Suppose that T satisfies (1.13), (1.33) and (1.36), then

$$(1.37) \qquad J(H_\Psi, d_{X,T}) = \int_0^\infty H_\Psi(\log N(K, d_{X,T}; \varepsilon))d\varepsilon < \infty$$

implies that $\{X(t)\}_{t\in K}$ has a version with continuous sample paths a.s. whereas either

$$(1.38) \qquad J(H_\Psi, d_{X,p}) = \int_0^\infty H_\Psi(\log N(K, d_{X,p}; \varepsilon))d\varepsilon = \infty$$

or

$$(1.39) \qquad \int_0^\infty (\log N(K, d_{X,T}; \varepsilon))^{1/q}d\varepsilon = \infty$$

implies that $\sup_{t\in K} |X(t)| = \infty$ a.s.

II.) Suppose that T satisfies (1.13), (1.34) and (1.35), then if either

$$(1.40) \qquad J(H_\Psi, d_{X,p}) = \int_0^\infty H_\Psi(\log N(K, d_{X,p}; \varepsilon))d\varepsilon < \infty$$

or

$$(1.41) \qquad \int_0^\infty (\log N(K, d_{X,T}; \varepsilon))^{1/q}d\varepsilon < \infty,$$

$\{X(t)\}_{t\in K}$ has a version with continuous sample paths a.s. whereas if

$$(1.42) \qquad J(H_\Psi, d_{X,T}) = \int_0^\infty H_\Psi(\log N(K, d_{X,T}; \varepsilon))d\varepsilon = \infty$$

$\sup_{t\in K} |X(t)| = \infty$ a.s.

If in this Corollary all the H_Ψ are replaced by H_T it would simply be a restatement of Theorem 1.1. However since $\Psi(x) \sim T(x)$ for x sufficiently large we see that $H_\Psi(x) \sim H_T(x)$ for x sufficiently

large. Therefore these results depend only on $\Psi(x)$ for x large as one

would expect (and also, of course, on the measure m which enters into

the determination of $d_{X,T}$.)

Theorem 1.2 is not completely analagous to Theorem 1.1. In (1.11)

for example η and Φ are interchangeable and thus (1.12) can be

interpreted two ways as we did in (1.38) and (1.39). However the domains

of ν and φ in (1.25) are different and consequently ν and φ are not

interchangeable in (1.25), that is, φ always gives rise to the norm and

ν determines the functions H_ν. Nevertheless it is still useful to

clarify Theorem 1.2 by considering some specific conditions on the

functions φ. Let φ be as defined in the paragraph containing

(1.19)-(1.23) and in addition suppose that it is regularly varying at zero

with index $1 < p < 2$. Thus we can write

$$(1.43) \qquad\qquad \phi(x) = x^p L(x)$$

where L satisfies $\lim\limits_{y \to 0} \frac{L(xy)}{L(y)} = 1$, $\forall 0 < x \leq 1$, i.e. L is a slowly varying

function at zero. We will consider the following conditions on φ:

$$(1.44) \qquad\qquad \phi(xy) \geq C\, x^p\, \phi(y) \qquad\qquad y \geq 0, \quad 0 \leq x \leq 1$$

$$(1.45) \qquad\qquad \phi(xy) \geq C\, \phi(y)\, \phi(y) \qquad\qquad y \geq 0, \quad 0 \leq x \leq 1$$

$$(1.46) \qquad\qquad \phi(xy) \leq c\, x^p\, \phi(y) \qquad\qquad y \geq 0, \quad 0 \leq x \leq 1$$

$$(1.47) \qquad\qquad \phi(xy) \leq c\, \phi(x)\, \phi(y) \qquad\qquad y \geq 0, \quad 0 \leq x \leq 1$$

Here C and c are constants not necessarily the same in each

appearance. Note that if L(x) is increasing (decreasing) in (1.43) then

(1.46) ((1.44)) holds. On the other hand if $\phi(x) = (ax)^p (\log \frac{1}{ax})^\beta$ then

(1.47) ((1.45)) holds if $\beta \geq 0$ ($\beta \leq 0$) if $a > 0$ is taken to be

sufficiently small. Clearly if $\phi(x) = x^p$ all the inequalities

(1.44)-(1.47) can be taken to be equalities with $C = c = 1$.

The next Corollary is obtained from Theorem 1.2 by taking $\nu(x) =$ $\phi(x)$ (in (1.25) and (1.30)) or $\nu(x) = x^p$ in an obvious way depending on which of the conditions (1.44)–(1.47) is satisfied.

__Corollary 1.4:__ Let $\{X(t)\}_{t\varepsilon K}$ be as given in (1.18). Assume that $\psi(x) \sim \phi(x)$ for x near zero where $\phi(x)$, $\phi(0) = 0$, is a convex function which is regularly varying at zero with index $1 < p < 2$ and let $\frac{1}{p} + \frac{1}{q} = 1$.

I.) Suppose that ϕ satisfies (1.27), (1.28), (1.44) and (1.47), then

$$(1.48) \qquad \int_0^\infty (\log N(K,d_{Y,\phi};\varepsilon))^{1/q} d\varepsilon < \infty$$

implies that the series representing $\{Y(t)\}_{t\varepsilon K}$ given in (1.18) converges uniformly a.s. whereas if

$$(1.49) \qquad J(\overline{H}_\psi, d_{Y,\phi}) = \int_0^\infty \overline{H}_\psi(\log N(K,d_{Y,\phi};\varepsilon)) d\varepsilon = \infty$$

then $\sup\limits_{t\varepsilon K} |Y(t)| = \infty$ a.s.

II.) Suppose that ϕ satisfies (1.27), (1.28), (1.45) and (1.46), then

$$(1.50) \qquad J(\overline{H}_\psi, d_{Y,\phi}) = \int_0^\infty \overline{H}_\psi(\log N(K,d_{Y,\phi};\varepsilon)) d\varepsilon < \infty$$

implies that the series representing $\{Y(t)\}_{t\varepsilon K}$ converges uniformly a.s. whereas if

$$(1.51) \qquad \int_0^\infty (\log N(K,d_{Y,\phi};\varepsilon))^{1/q} d\varepsilon = \infty$$

then $\sup\limits_{t\varepsilon K} |X(t)| = \infty$ a.s.

If in this Corollary ψ is replaced by ϕ it would simply be a restatement of Theorem 1.2. However since $\psi \sim \phi$ near zero we see that $\overline{H}_\psi(y) \sim \overline{H}_\phi(x)$ for x sufficiently large. Thus in this Corollary the results depend only on $\psi(x)$ for x near zero as one would expect (and also, of course, on the sequence $\{a_\gamma\}_{\gamma\varepsilon A}$).

Note that the necessary conditions in these Corollaries hold under more general conditions than those given. Conditions (1.13), (1.27) and (1.28) are not required in the proof of necessity.

In Corollary 1.3 if $\Psi(u) = u^p$, $1 < p < 2$, we get the main result of [MP2], i.e.

(1.52) $\int_0^\infty \left(\log N(K, d_{X,p}; \varepsilon)\right)^{1/q} d\varepsilon < \infty$, $\frac{1}{q} + \frac{1}{p} = 1$,

is necessary and sufficient for $\{X(t)\}_{t \in K}$ to have a version with continuous sample paths, where we recall that

$$d_{X,p}(s,t) = \left(E_m |\gamma(s) - \gamma(t)|^p\right)^{1/p} .$$

Of course, in this case, we have p-stable processes. Similarly in Corollary 1.4 if $\psi(u) = u^p$, $1 < p < 2$, we get that

(1.53) $\int_0^\infty \left(\log N(K, d_{Y,p}; \varepsilon)\right)^{1/q} d\varepsilon < \infty$, $\frac{1}{q} + \frac{1}{p} = 1$,

is necessary and sufficient for the series in (1.18) to converge uniformly a.s. In this case recall that

$$d_{Y,p}(s,t) = \left(\sum_{\gamma \in A} |a_\gamma|^p |\gamma(s) - \gamma(t)|^p\right)^{1/p}$$

and the random variable ξ is p-stable. Notice also, that in the case of p-stable processes, Theorem 1.2 is a special case of Theorem 1.1 since in Theorem 1.1 if the measure m is supported on the countable set $A \subset \Gamma$ and is such that $m(\{\gamma\}) = |a_\gamma|^p$, $\gamma \in A$, we see that $X(t)$ is equal in distribution to $\sum_{\gamma \in A} a_\gamma \xi_\gamma \gamma(t)$, where $\{\xi_\gamma\}$ are i.i.d. p-stable. On the other hand, Theorem 1.2 is not a discrete version of Theorem 1.1 when $\Psi(u)$ and $\psi(u)$ are different from u^p. Indeed, the two Theorems are very different. Theorem 1.1 depends on $\Psi(u)$ for u large whereas Theorem 1.2 depends on $\psi(u)$ for u near zero. Moreover, even if the measure m is supported on $A \subset \Gamma$, $\{X(t)\}_{t \in K}$ is not expressible as a series of the form (1.18). (To be more explicit, if the support of m is countable

$\{X(t)\}_{t\in K}$ is always equal in distribution to a random Fourier series with independent coefficients but the coefficients are not of the form $\{b_\gamma \xi_\gamma\}$ where $\{b_\gamma\}$ are complex numbers and $\{\xi_\gamma\}$ are i.i.d.).

As a further elaboration of Corollaries 1.3 and 1.4 we give the following more specific results. They show how close our necessary and sufficient conditions can be.

Corollary 1.5: Let $\{X(t)\}_{t\in K}$ be a $\mathbb{C}^K$ valued stochastic process with characteristic functional given by (1.3) or (1.5) with $\Psi(u) \sim u^p (\log u)^\beta$, $1 < p < 2$, for u sufficiently large. Let $\Phi: R^+ \rightarrow R^+$ be a strictly increasing convex function satisfying

$$(1.54) \qquad\qquad \Phi(u) \sim u^p (\log (e+u))^\beta, \quad u \geq u_0$$

for some $u_0 \geq 0$. For $s, t \in K$, let $d_{X,\Phi}(s,t)$ and $d_{X,p}(s,t)$ be given as in (1.6) and let

$$(1.55) \qquad\qquad H(u) = \begin{cases} u^{1/q}(\log u)^{\beta/p}, & u \geq 1, \frac{1}{p} + \frac{1}{q} = 1 \\ 0, & 0 \leq u < 1 \end{cases}.$$

I.) Let $\beta \geq 0$. Then

$$(1.56) \qquad\qquad \int_0^\infty H(\log N(K, d_{X,\Phi}; \varepsilon) < \infty$$

implies $\{X(t)\}_{t\in K}$ has a version with continuous paths a.s., whereas either

$$(1.57) \qquad\qquad \int_0^\infty (\log N(K, d_{X,\Phi}; \varepsilon))^{1/q} d\varepsilon = \infty$$

or

$$(1.58) \qquad\qquad \int_0^\infty H(\log N(K, d_{X,p}; \varepsilon)) d\varepsilon = \infty$$

implies that $\sup_{t\in K} |X(t)|$ is unbounded a.s.

II.) Let $\beta \leq 0$. Then either

(1.59) $\int_0^\infty (\log N(K,d_{X,\phi};\epsilon))^{1/q} \, d\epsilon < \infty$

or

(1.60) $\int_0^\infty H(\log N(K,d_{X,p};\epsilon)) d\epsilon < \infty$

implies that $\{X(t)\}_{t \in K}$ has a version with continuous paths whereas

(1.61) $\int_0^\infty H(\log N(K,d_{X,\phi};\epsilon)) d\epsilon = \infty$

implies $\sup_{t \in K} |X(t)|$ is unbounded a.s.

<u>Corollary 1.6</u>: Let $\psi(\lambda) \sim \lambda^p(\log 1/\lambda)^\beta$, $1 < p < 2$, for $\lambda > 0$
sufficiently close to zero; let ξ be given as in (1.17) and consider the
random Fourier series $\{Y(t)\}_{t \in K}$ as defined in (1.18) where $\{\xi_\gamma\}_{\gamma \in A}$ are
i.i.d copies of ξ. Let $\phi: R^+ \to R^+$ be a strictly increasing convex
function satisfying

$$\phi(\lambda) \sim \lambda^p(\log(e + 1/\lambda))^\beta, \quad \lambda \leq \lambda_0$$

for some $\lambda_0 > 0$. For s, t $\in$ K let $d_{Y,\phi}(s,t)$ be as in (1.21) for this
function ϕ and let

(1.62) $H(x) = \dfrac{x^{1/q}}{(\log x)^{\beta/p}}, \quad x \geq e, \quad \dfrac{1}{p} + \dfrac{1}{q} = 1 .$

I.) If $\beta \geq 0$, then

(1.63) $\int_0^\infty (\log N(K,d_{Y,\phi};\epsilon))^{1/q} \, d\epsilon < \infty$

implies that $\{Y(t)\}_{t \in K}$ has a version with continuous paths a.s., whereas

(1.64) $\int_0^\infty H(\log N(K,d_{Y,\phi};\epsilon)) d\epsilon = \infty$

implies that $\sup_{t \in k} |Y(t)|$ is unbounded a.s.

II.) If $\beta \leq 0$, then

(1.65) $\int_0^\infty H(\log N(K, d_{Y,\phi}; \varepsilon)) d\varepsilon < \infty$

implies that $\{Y(t)\}_{t \in K}$ has a version with conitnuous paths a.s., whereas

(1.66) $\int_0^\infty (\log N(K, d_{Y,\phi}; \varepsilon))^{1/q} d\varepsilon = \infty$

implies that $\sup_{t \in K} |Y(t)|$ is unbounded a.s.

Note that the existence of processes for which Ψ and ψ have the asymptotic properties described in Corollaries 1.3 and 1.4 follow from Tauberian theorems relating the Levy transform to its corresponding Levy measure (see e.g. Theorem 4, [BT]) and the characteristic function to its corresponding distribution function.

So far we have concentrated on a specific ξ-radial process as determined by its Levy measure Ψ and have obtained results on sample path continuity which are valid for all measures m. Likewise in the random Fourier series case we pick a specific i.i.d. sequence $\{\xi_\gamma\}_{\gamma \in A}$ as determined by its characteristic function ψ and give results which are valid for all sequences of coefficients $\{a_\gamma\}_{\gamma \in A}$. Unfortunately, except for the p-stable case, $1 < p \leq 2$, which is already known, none of our results are both necessary and sufficient. For specific examples in which the measure m or the sequence $\{a_\gamma\}_{\gamma \in A}$ is given precisely more exact results can be obtained. This is done in Chapter 5 using a technique of Cuzick and Lai [CL] for sufficiency and an extension of a classical result of Paley and Zygmund [PZ] for necessity.

We have two theorems along this line which are worth mentioning here because they are extensions of an elegant classical result of Salem and Zygmund. For background consider

(1.67) $Z(t) = \sum_{k=1}^\infty a_k \theta_k e^{ikt}, \quad t \in [0, 2\pi]$

where $\{\theta_k\}$ are i.i.d. p-stable, $1 < p \leq 2$, and define

$$s_j = \left(\sum_{k=2^j}^{2^{j+1}-1} |a_k|^p \right)^{1/p}.$$

Then in the case when s_j is non-increasing

(1.68) $\sum_j s_j < \infty$ iff $Z(t)$ converges uniformly a.s.

When $p = 2$ this result is due to Salem and Zygmund with a contribution
by Kahane (see [K], Chapter 8, Section 5 for more details). In this case
the result remains valid with $\{\theta_k\}$ replaced by any i.i.d. symmetric
sequence with finite second moments. The case $1 < p < 2$ is given in
Lemma 2.3, [MP3]. As we have already pointed out p-stable series can be
represented as ξ-radial processes or as random Fourier series.
However when $\{\theta_k\}$ in (1.67) is not p-stable or when $\Psi(\lambda) \neq |\lambda|^p$ in
(1.3) or (1.5) these two classes of processes are different. Nevertheless
in each of these cases we get an extension of (1.68). Theorem 1.7 deals
with ξ-radial processes and Theorem 1.8 with random Fourier series.

<u>Theorem 1.7</u>: Let $G = [0, 2\pi]$ so that $\Gamma = \{e^{ikt}, k \in \mathbb{N}\}$. Let
$\{X(t)\}_{t \in G}$ be a $\mathbb{C}^G$ valued stochastic process with characteristic
functional given by (1.5) where $\Psi(u)$ is regularly varying at infinity
with index $1 < p < 2$. Define

(1.69) $m_i = m\{e^{ikt}: 2^i \leq k < 2^{i+1}\}$

where m is the measure in (1.5). Then if $\Psi^{-1}\left(\dfrac{1}{m_i}\right)$ is non-decreasing in i

(1.70) $\displaystyle\sum_{i=1}^{\infty} \frac{1}{\Psi^{-1}\left(\frac{1}{m_i}\right)} < \infty$

is necessary and sufficent for $\{X(t)\}_{t \in G}$ to have a version with continuous sample paths. (We shall explain in Chapter 5 how we define ψ^{-1} when ψ is not non-decreasing at infinity.)

Theorem 1.8: Let $G = [0, 2\pi]$ so that $\Gamma = \{e^{ikt}, \ k \in \mathbb{N}\}$. Let $\psi(\lambda)$, $\psi(1) = 1$, be a regularly varying function at zero with index $1 < p < 2$. Let ξ be as given in (1.17) and consider

$$(1.71) \qquad Y(t) = \sum_{k=0}^{\infty} a_k \xi_k e^{ikt}, \quad t \in [0, 2\pi].$$

Define

$$(1.72) \qquad \| \{a_k\}_{k=2^n}^{2^{n+1}-1} \|_\psi = \inf \ \{\lambda > 0: \ \sum_{k=2^n}^{2^{n+1}-1} \psi\left(\frac{a_k}{\lambda}\right) \leq 1\}.$$

Then if $\| \{a_k\}_{k=2^n}^{2^{n+1}-1} \|_\psi$ is non-increasing in n

$$(1.73) \qquad \sum_{n=1}^{\infty} \| \{a_k\}_{k=2^n}^{2^{n+1}-1} \|_\psi < \infty$$

is necessary and sufficient for $\{X(t)\}_{t \in G}$ to converge uniformly a.s. (We shall explain in Chapter 5 how we define $\| \ \|_\psi$ when ψ is not non-decreasing at zero.)

In Chapter 6 we consider ξ-radial processes for which the Levy measure Ψ is regularly varying at infinity with index 1 or 2 and random Fourier series for which the characteristic function is regularly varying at the origin with index 1 or 2. The results that are obtained are somewhat more complicated because of the failure of certain Tauberian theorems that are used in the proofs of Theorems 1.1 and 1.2. The purpose of these examples and extensions is to enable us to get a better idea of whether there is a single integral condition depending on the metric entropy which gives necessary and sufficient conditions for a ξ-radial process (in which Ψ is fixed and m is arbitrary) or for a random

Fourier series (in which ψ is fixed and $\{a_\gamma\}_{\gamma\epsilon A}$ is arbitrary) to have a version with continuous sample paths. We know that such integral conditions exist for p-stable processes, $1 < p \leq 2$. They are given in (1.52) and (1.53) and are also valid when $p = 2$ since in this case the processes are Gaussian.

Based on Theorems 1.1 and the many examples given in Chapters 5 and 6 we make the following conjecture for ξ-radial processes:

<u>Conjecture</u>: Let $\{X(t)\}_{t\epsilon K}$ be an $\mathbb{R}^K(\mathbb{C}^K)$ valued stochastic process with characteristic functional given by (1.3) ((1.5)) and assume that $\Psi(x)$ is regularly varying at infinity with index $0 < p \leq 2$. Then

(1.74) $J(H_\Psi, d_{X,p}) = \int_0^\infty H_\Psi(\log N(K, d_{X,p}; \epsilon)) d\epsilon < \infty$

is necessary and sufficient for $\{X(t)\}_{t\epsilon K}$ to have a version with continuous sample paths, where for $s, t \epsilon K$

(1.75) $d_{X,p}(s,t) = (E_m |X(t) - X(s)|^p)^{1/p}$

and

$$H_\Psi(x) = \begin{cases} \int_1^x \dfrac{ds}{\Psi^{-1}(s)}, & x \geq 1 \\ \\ 0 & 0 \leq x < 1 \end{cases}$$

(Note that $H_\Psi(x) \sim \dfrac{x}{\Psi^{-1}(x)}$ if $\Psi(x)$ is regularly varying with index $1 < p < 2$. Also $\{X(t)\}_{t\epsilon K}$ always has a version with continuous sample paths if $\Psi(x)$ is regularly varying with index $0 < p < 1$.)

This conjecture is consistent with Theorem 1.7 and the other relevant examples of Chapters 5 and 6. Here are several reasons why this conjecture is appealling. It is known that if

(1.76) $\left| \int_0^\infty (1 \wedge t) d\tau [t, \infty) \right| < \infty$

then $\{X(t)\}_{t\in K}$ has continuous paths for all measures m, (see Lemma 7.2).
By Theorem 4, [BT], (1.76) holds if and only if H_ψ is bounded.
Obviously, if H_ψ is bounded the integral in (1.74) is finite. Thus the
conjectured result gives the trivial cases determined by (1.76). At the
other extreme, i.e. when Ψ is regularly varying with index 2, we see by
(6.30) that

(1.77) $\Psi(x) = o(x^2)$ as $x \to \infty$.

Therefore

(1.78) $H_\psi(x) = o(x^{1/2})$ as $x \to \infty$.

Examples show that $H_\psi(x)$ can get arbitrarily close to $x^{1/2}$ as long as
(1.78) is preserved. Thus $J(H_\psi, d_{X,p})$ has as its "boundary" the
condition for continuity of stationary Gaussian processes just as
infinitely divisible random variables have as their boundary the Gaussian
random variable.

At first glance, it seems pecular that the metric in (1.75) only
depends on p, the index of regular variation of Ψ. However, this seems
to be the case. The examples in Corollary 5.4 show that $J(H_\psi, d_{X,\psi})$, (see
(1.37) and note that one can take $T = \Psi$ if Ψ is comparable to a convex
function at infinity), is not a necessary and sufficient condition for the
continuity of $\{X(t)\}_{t\in K}$ and the remarks following Theorem 6.1 show that
the integral condition in (1.41) is not a necessary and sufficient
condition for continuity of $\{X(t)\}_{t\in K}$.

One might ask, what kind of integral condition could give necessary
and sufficient conditions for continuity when Ψ is not regularly varying
at infinity? Perhaps the answer is none, at least none of the form of
(1.7) with a fixed function H and a fixed metric d. (It is easy to
give necessary and sufficient conditions for continuity in terms of a
random metric since one can simply require that the marginal subgaussian

processes formed from (1.18), or (2.21) with $\{h_j\}$ a Rademacher sequence,
are continuous a.s. and apply Theorem 1.1 Chapter I of [MP1] to these
marginal processes.) In any case we know that real valued infinitely
divisible random variables with regularly varying probability
distributions play the special role of being in the domain of attraction
of the stables. Perhaps the class of ξ-radial processes with regularly
varying Levy transforms is similarly special.

Results from Theorems 1.1 and 6.6 prove parts of the conjecture. We
see from Theorem 1.1, I.) that $J(H_\Psi, d_{X,p}) = \infty$ implies $\sup_{t \in K} |X(t)| = \infty$
a.s. when $\Psi(xy) \geq c \, x^p \, \Psi(y)$ for $x \geq x_1$, $y \geq y_1$ for some x_1, y_1
sufficiently large. This is valid for $1 \leq p < 2$. When $p = 2$ this
inequality can't hold because if it did Ψ wouldn't satisfy (1.77). For
$p < 1$, $J(H_\Psi, d_{X,p})$ is always finite. Also, under addition regularity
conditions on Ψ, (it must be comparable at infinity to a function $\eta(x)$
that satisfies (1.13) and when $p = 2$ the conditions of Theorem 6.6 must
hold), we see that $J(H_\Psi, d_{X,p}) < \infty$ implies that $\{X(t)\}_{t \in K}$ has a version
with continuous sample paths when $\Psi(xy) \leq C \, x^p \, \Psi(y)$, $\forall x, y \geq 1$ and
$1 < p \leq 2$. Our methods don't work to give sufficient conditions in the
case $p = 1$. In fact, this case is not covered in [MP2] in which stable
processes are considered.

For random Fourier series the situation is simpler. There is no
integral condition of the form of (1.7), with a fixed function H and
metric d, which gives necessary and sufficient conditions for continuity
of random Fourier series except when ξ in (1.17) is p stable. (The
1-stable case is not yet resolved.) Consider (1.48) and (1.50) of
Corollary 1.4. The examples in Corollary 5.5 are consistent with (1.48)
being a necessary and sufficient condition for continuity and show that
(1.50) can not be. On the other hand at the end of Chapter 5 we study
examples of lacunary random Fourier series which are consistent with

(1.50) being a necessary and sufficient condition for continuity and show
that (1.48) can not be. Therefore, at least for ψ as in Corollary 5.5,
the gap that exists in Corollary 1.4 can not be closed. It appears
unlikely that it could be for even more irregular functions ψ. Of course
when $\psi(x) = |x|^p$, (1.48) and (1.50) are equivalent. (In [LM], in which
Gaussian Fourier quadratic forms are considered, we also find that no
condition of the form (1.7) determines continuity.)

Since no entropy condition of the form of (1.7) determines continuity
for random Fourier series of the form of (1.18) which are not p-stable, if
the conjecture is correct, it will indicate that ξ-radial processes are
the more natural class of stochastic processes. That is, that the
reasonable generalization of random Fourier series with p-stable
coefficients is to ξ-radial processes not to other random Fourier series
of the form (1.18) in which the i.i.d. sequence $\{\xi_\gamma\}_{\gamma \in A}$ is not p-stable.

In Chapter 2 we develop a representation for ξ-radial infinitely
divisible stochastic processes originated by Le Page [LeP]. Some of the
results are new and interesting even for symmetric real valued infinitely
divisible random variables. The representation enables us to consider the
processes defined in (1.3) and (1.5) as random Fourier series with
sign-invariant terms and permits us to study them with the methods used in
[MP1] and [MP2].

In Chapter 3 we prove the necessity part of Theorems 1.1 and 1.2 and
in Chapter 4 we prove the sufficiency part. In both of these sections we
obtain interesting relationships between random and non-random metrics
which may prove useful in other contexts. (See e.g. Lemma 3.2 and Theorem
4.6 and Corollary 4.7). Note that Corollaries 1.3, 1.4, 1.5 and 1.6 need
no proof. They are simply special cases of Theorems 1.1 and 1.2.

Chapter 5 includes the proofs of Theorems 1.7 and 1.8 and many other
examples.

In Chapter 6 we consider the case when Ψ and ψ are regularly varying with index $p = 1$ and $p = 2$.

In Chapter 7 we return to a popular theme in the study of probability on Banach spaces. Given that a ξ-radial process (i.e. a process defined by (1.3) or (1.5)) has continuous sample paths we ask what is the distribution of its sup-norm. This question can be resolved in many cases, including those considered in Corollary 1.5. We observe an interesting property of ξ-radial processes. That is, whereas the continuity of the process depends on $\Psi(|\lambda|)$ for $|\lambda|$ large, the distribution of the sup-norm of a continuous process is determined by $\Psi(|\lambda|)$ for $|\lambda|$ near zero. For example if $\Psi(|\lambda|) \sim \lambda^p$ as $\lambda \downarrow 0$ and $\Psi(|\lambda|) \sim \lambda^{p'}$ as $\lambda \to \infty$ the continuity of the corresponding real or complex valued ξ-radial process is the same as that of a p'-stable process but when the process is continuous the distribution of its sup-norm is the same as that of a p-stable process. The answer to this question for the processes defined in (1.18) is also discussed in Section 7.

The formulation of Theorem 1.1 was begun in collaboration with Gilles Pisier as a natural continuation of [MP2]. I very much appreciate his contribution to this work. I am also grateful to Evarist Giné and Joel Zinn for several helpful discussions about this work and to Xavier Fernique for showing me how to describe class Θ.

2. REPRESENTING ξ-RADIAL PROCESSES

A symmetric real valued infinitely divisible random variable ξ is one with characteristic function given by

$$(2.1) \qquad Ee^{i\lambda\xi} = \exp -\Psi(|\lambda|)$$

where

$$(2.2) \qquad \Psi(|\lambda|) = \int_0^\infty (\cos \lambda t - 1)d\tau[t,\infty)$$

and τ is a Levy measure, i.e. τ is a positive measure on $R^+/\{0\}$ which satisfies

$$(2.3) \qquad \left| \int_0^\infty (1 \wedge t^2)d\tau[t,\infty) \right| < \infty.$$

We will call a function Ψ associated with a Levy measure τ, as in (2.2), the Levy transform of τ.

We will also be interested in those complex valued infinitely divisible random variables $\tilde{\xi}$ which satisfy, $\forall z \in \mathbb{C}$,

$$(2.4) \qquad E \exp i \, \text{Re}(\overline{z}\tilde{\xi}) = \exp -\Psi(|z|),$$

for some Levy transform Ψ. It is clear that not all Levy transforms are such that (2.4) is satisfied for all $z \in \mathbb{C}$. However, as we will later show, there are many for which such a definition is possible.

Let T be a set. We will denote by $R^{(T)}$ (resp $C^{(T)}$) the space of all finitely supported families $(\alpha(t))_{t\in T}$ of real (resp. complex) numbers. We will say that a real (resp. complex) valued stochastic processes is <u>ξ-radial</u> if there exists a probability measure m on the real (resp. complex) valued functions $\beta(t)$, $t \in T$ such that

$\sup_{t \varepsilon T} |\beta(t)| = 1$, equipped with the cylindrical σ-algebra, such that
$\forall \alpha \varepsilon \mathbb{R}^{(T)}$

(2.5) $E \exp i \sum_{t \varepsilon T} \alpha(t)X(t) = \exp - \int \Psi\left(\left|\sum_{t \varepsilon T} \alpha(t)\beta(t)\right|\right)m(d\beta),$

(resp. $\forall \alpha \varepsilon \mathbb{C}^{(T)}$)

(2.6) $E \exp i \operatorname{Re}\left|\sum_{t \varepsilon T} \overline{\alpha(t)}X(t)\right| = \exp - \int \Psi\left(\left|\sum_{t \varepsilon T} \overline{\alpha(t)}\beta(t)\right|\right)m(d\beta),$

where the connection between ξ and Ψ is given by (2.1). The
measures m in (2.5) and (2.6) are called the spectral measures of the
corresponding processes.

Now let T be a locally compact Abelian group G with dual group
Γ. Γ is called the character group of G, i.e. $\gamma \varepsilon \Gamma$ is a
continuous complex valued function such that $\forall s, t \varepsilon G$, $\gamma(t) = 1$
and $\gamma(s)\gamma(t) = \gamma(s+t)$. A real (resp. complex) valued ξ-radial process
$(X(t))_{t \varepsilon G}$ will be called <u>strongly stationary</u> if it admits a
representation as in (2.5) (resp. (2.6)) where the spectral measure m
is a probability measure supported on Γ.

A strongly stationary ξ-radial process is stationary, since for
$(X(t))_{t \varepsilon T}$ real $t_1, \ldots, t_n \varepsilon T$ and $t_1+s, \ldots, t_n+s \varepsilon T$,

$$E \exp i \sum_{j=1}^{n} \alpha_j X(t_j+s) = E \exp i \sum_{j=1}^{n} \alpha_j X(t_j).$$

This follows from (2.5). In the complex case we say that $(X(t))_{t \varepsilon T}$
is stationary if for every $t_1, \ldots, t_n \varepsilon T$ the $2n$ dimensional real
sequence $(\operatorname{Re}X(t_1), \operatorname{Im} X(t_1), \ldots, \operatorname{Re} X(t_n), \operatorname{Im} X(t_n))$ is stationary.
Clearly the complex valued ξ-radial processes defined by (2.6) are also
stationary. When $\Psi(|\lambda|) = C|\lambda|^p$, $0 < p < 2$, for some constant C,
(2.5) and (2.6) define real and complex valued strongly stationary
p-stable processes which were studied in [MP2] in which we showed that
not all stationary p-stable processes are strongly stationary contrary
to the situation for Gaussian processes.

A major step in our work is the representation of the processes defined in (2.5) and (2.6) as series which are marginally independent and, in some cases, which are marginally Gaussian processes. That this can be done is essentially due to LePage [LeP]. The proofs that are given here are based in part on unpublished notes of E. Giné and J. Zinn. For our purposes, everything will follow from appropriate results in one dimension.

Let τ be a Levy measure and h a symmetric real valued random variable. Define

$$(2.7) \qquad \tau_h[t,\infty) = E_h\left(\tau\left[\frac{t}{|h|}, \infty\right)\right), \quad t > 0$$

where E_h denotes expectation with respect to the law of h. We are only interested in those cases in which τ_h is itself a Levy measure, i.e. τ_h satisfies (2.3). (Whether τ_h is a Levy measure depends both on τ and h but if $E_h h^2 < \infty$ then τ_h is a Levy measure. This is because $E_h(1 \wedge h^2 v^2) \leq 1 \wedge (E_h h^2)v^2$.) If τ_h is a Levy measure we denote by Ψ_h the Levy transform of τ_h. Clearly

$$(2.8) \qquad E_h \Psi(|\lambda h|) = \Psi_h(|\lambda|) , \quad \forall \lambda \text{ real}$$

where Ψ is the Levy transform of τ and (2.7) is finite if and only if (2.8) is finite. When (2.7) defines a Levy measure τ_h with corresponding Levy transform Ψ_h we denote the corresponding symmetric infinitely divisible random variable defined by (2.1) - (2.3) by ξ_h.

Let X be a non-negative random variable satisfying $P[X > \lambda] = e^{-\lambda}$. Let $\{X_j\}_{j=1}^{\infty}$ be i.i.d copies of X and define

$$(2.9) \qquad \Gamma_j = X_1 + \dots + X_j.$$

By [F] p. 10

$$(2.10) \qquad P[\Gamma_j < x] = \int_0^x \frac{x^{j-1}}{(j-1)!} e^{-x} dx .$$

Lemma 2.1: Let ξ, Ψ and τ be defined by (2.1) – (2.3) and let h be a symmetric real valued random variable, independent of ξ satisfying (2.7) and (2.8) for all $\lambda \geq 0$ such that τ_h is a Levy measure. Then ξ_h is a symmetric infinitely divisible random variable with Levy transform τ_h and furthermore

(2.11)
$$\xi_h \overset{\mathcal{D}}{=} \sum_{j=1}^{\infty} F^{-1}(\Gamma_j) h_j$$

where "$\mathcal{D}$" denotes equal in distribution and $\{h_j\}_{j=1}^{\infty}$ are i.i.d. copies of h, $F^{-1}(t) = \sup\{u \colon \tau[u,\infty) > t\}$, and the series converges a.s.

Proof: That ξ_h is an infinitely divisible random variable with Levy measure τ_h and Levy transform Ψ_h follows from the definition of ξ_h.   Now let $K_n = \#\{j \colon \Gamma_j \leq n\}$. Then $P[K_n = k] = e^{-n}\dfrac{n^k}{k!}$, (see e.g. [F], pg. 11). Let U be a real valued uniformly distributed random variable on $[0,1]$ and let $\{U_i\}_{i=1}^{k}$ be i.i.d. copies of U. Let $\{U_{(i)}\}_{i=1}^{k}$ be a non-decreasing rearrangement of $\{U_i\}_{i=1}^{k}$. By the same proof as that of Prop. 13.15, [Bl] one can show that for $\forall a > 0$

(2.12) $L\left(\left(\dfrac{\Gamma_1}{a}, \ldots, \dfrac{\Gamma_k}{a}\right)\Big|\Gamma_k \leq a, \Gamma_{k+1} > a\right) = L\left(\left(U_{(1)}, \ldots, U_{(k)}\right)\right),$

where we use the notation $L(X)$ to denote the probability measure associated with the random variable X. We now consider the random sum

$$S_n = \sum_{j=1}^{K_n} F^{-1}(\Gamma_j) h_j \ .$$

For each k and $\{U_i\}_{i=1}^{k}$ independent of $\{X_j\}_{j=1}^{\infty}$ and $\{h_j\}_{j=1}^{\infty}$ we have

$$L\left(S_n\Big| K_n = k\right) = L\left(\sum_{j=1}^{k} F^{-1}(nU_{(j)}) h_j\Big| K_n = k\right)$$

$$= L\left(\sum_{j=1}^{k} F^{-1}(nU_j) h_j\Big| K_n = k\right).$$

Therefore the characteristic function of S_n, denoted by $\hat{S}_n$, is

(2.13)
$$\hat{S}_n(\lambda) = e^{-n} \sum_{k=0}^{\infty} \frac{n^k}{k!} E \exp\left[i\lambda \sum_{j=1}^{k} F^{-1}(nU_j)h_j \right]$$

$$= e^{n(\hat{\phi}_n(\lambda)-1)} = e^{-\Psi_n(|\lambda|)}$$

where $\hat{\phi}_n$ is the characteristic function of $F^{-1}(nU)h$ and Ψ_n is the Levy transform of $m \times \tau_n$, where $\tau_n(A) = \tau([F^{-1}(n), \infty) \cap A)$ for all $A \in R^+$, $t > 0$ and $m = L(h)$. Clearly $m \times \tau_n \to m \times \tau$ as $n \to \infty$ and consequently $\Psi_n(|\lambda|) \uparrow \Psi_h(|\lambda|)$. This shows that S_n converges in distribution to ξ_h and this implies that $\sum_{j=1}^{\infty} F^{-1}(\Gamma_j)h_j$ converges in distribution and is equal in distribution to ξ_h. Furthermore since (2.11) is symmetric and marginally independent we know also that the series converges a.s.

When the Levy measure τ satisfies the stronger integrability condition

(2.14)
$$\left| \int_0^{\infty} (1 \wedge t) d\tau[t,\infty) \right| < \infty$$

we can define a strictly positive infinitely divisible random variable ξ by

(2.15)
$$E^{i\lambda\xi} = \exp[-\Psi(|\lambda|)]$$

where

(2.16)
$$\Psi(|\lambda|) = \int_0^{\infty} (e^{i\lambda t} - 1) d\tau[t,\infty) .$$

We use the same notation as in (2.1), (2.2) and (2.3) because when dealing with positive infinitely divisible random variables all the other relevant quantities, τ_h, Ψ_h, ξ_h, etc, are exactly as in (2.7) and (2.8) except that Ψ is given by (2.16) instead of (2.2). The following is the analogue of Lemma 2.1 for positive infinitely divisible random variables. The proof is exactly the same as the proof of Lemma 2.1.

Lemma 2.2: Let ξ be defined by (2.14) - (2.16) and let h be a positive random variable independent of ξ satisfying (2.8) for all $\lambda \geq 0$ but with Ψ as given in (2.16). Assume that τ_h also satisfies (2.14) then ξ_h is an infinitely divisible random variable with Levy transform τ_h as given in (2.7). Furthermore

$$(2.17) \qquad \xi_h \overset{\mathcal{D}}{=} \sum_{j=1}^{\infty} F^{-1}(\Gamma_j)h_j$$

where $\{h_j\}_{j=1}^{\infty}$ are i.i.d. copies of h, F^{-1} is as defined in Lemma 2.1 and the series converges a.s.

We are now ready to obtain series representations of ξ-radial processes. To avoid confusion we will collect some of the previous definitions here. To begin with let Ψ be the Levy transform of a symmetric infinitely divisible random variable as defined in (2.1) and (2.2) and let τ be the associated Levy measure. As in Lemma 2.1 let

$$(2.18) \qquad F^{-1}(t) = \sup [u: \tau[u,\infty) > t] \ .$$

Let h be a symmetric real valued random variable and let Ψ_h and τ_h be as defined in (2.7) and (2.8). Under the assumption that τ_h is a Levy measure, consider the real and complex valued ξ-radial processes defined by

$$(2.19) \qquad E \exp i \sum_{t \in T} \alpha(t)X(t) = \exp - \int \Psi_h\left(\left|\sum_{t \in T} \alpha(t)\beta(t)\right|\right)m(d\beta)$$

and

$$(2.20) \qquad E \exp i \ \mathrm{Re}\left[\sum_{t \in T} \overline{\alpha(t)}X(t)\right] = \exp - \int \Psi_h\left(\left|\sum_{t \in T} \overline{\alpha(t)}\beta(t)\right|\right)m(d\beta)$$

where α and m and T are as given in (2.5) and (2.6). (Indeed (2.19) and (2.20) are precisely (2.5) and (2.6) with Ψ replaced by Ψ_h). Let $\{Y_j\}$ be a sequence of i.i.d. R^T (resp. C^T) valued random variables with probability distribution m and let $\{h_j\}$ be i.i.d.

copies of   h.   Let $\{\varepsilon_j\}$ be a Rademacher sequence (i.e. an i.i.d.
sequence of symmetric random variable each one taking the values ± 1) and
let $\{\theta_j\}$ be an i.i.d. sequence of random variables uniformly
distributed on $[0,2\pi]$.   Let $\{g_j\}$ and $\{g_j'\}$ be mutually independent
sequences of i.i.d. normal random variables with mean zero and equal
variance and let $\{\tilde{g}_j\}$ be an i.i.d. sequence consisting of copies of
$g_1 + ig_1'$.   Let $\{\Gamma_j\}$ be as given in (2.9).   We assume that all the
sequences, $\{Y_j\}$, $\{\Gamma_j\}$, $\{\varepsilon_j\}$, $\{\theta_j\}$, $\{h_j\}$, $\{\tilde{g}_j\}$ are independent of the
others.   In what follows we will occasionally denote the integral with
respect to the probability measure m   by E_m.

Lemma 2.3:   In the above notation let Ψ and τ be associated by (2.2)
where τ is a Levy measure and let $F^{-1}(t) = \sup\{u: \tau[u,\infty) > t\}$.
Assume that τ_h is a Levy measure with Levy transform Ψ_h.

Let $(X(t))_{t\epsilon T}$ be a real valued ξ-radial process as defined in
(2.19).   Then

(2.21) $$V(t) = \sum_{j=1}^{\infty} F^{-1}(\Gamma_j)h_j Y_j(t), \quad t \epsilon T ,$$

is equal in distribution to $(X(t))_{t\epsilon T}$.

Let $\{k_j\}$ be i.i.d. symmetric real valued random variables
independent of all the other random sequences.   Let $h = k_1 g_1$ and
assume that τ_h is a Levy measure.   Let $(X(t))_{t\epsilon T}$ be a complex valued
ξ-radial process as defined in (2.20).   Then

(2.22) $$Z(t) = \sum_{j=1}^{\infty} F^{-1}(\Gamma_j)k_j\tilde{g}_j Y_j(t), \quad t \epsilon T,$$

is equal in distribution to $(X(t))_{t\epsilon T}$.

Let $\{k_j\}$ be i.i.d. symmetric real valued random variables
independent of all the other random sequences.   Let $h = k_1 \cos \theta_1$ and
assume that τ_h is a Levy measure.   Let $(X(t))_{t\epsilon T}$ be a complex valued
ξ-radial process as defined in (2.20).   Then

(2.23) $W(t) = \sum\limits_{j=1}^{\infty} F^{-1}(\Gamma_j) k_j e^{i\theta_j} Y_j(t), \quad t \in T,$

is equal in distribution to $(X(t))_{t \in T}$.

Proof: We begin with the real case. We have

(2.24) $\sum\limits_{t \in T} \alpha(t) V(t) \overset{\mathcal{D}}{=} \sum\limits_{j=1}^{\infty} F^{-1}(\Gamma_j) h_j \Big(\sum\limits_{t \in T} \alpha(t) Y_j(t) \Big),$

where "$\mathcal{D}$" means equal in distribution. By Lemma 2.1, $\sum\limits_{t \in T} \alpha(t) V(t)$ is equal in distribution to a symmetric infinitely divisible random variable with Levy measure τ_ρ and Levy transform Ψ_ρ where $\rho = h\Big(\sum\limits_{t \in T} \alpha(t) Y_1(t) \Big)$. However, since h and $\Big(\sum\limits_{t \in T} \alpha(t) Y_1(t) \Big)$ are independent we have

(2.25) $\Psi_h(|\lambda|) = E_m \Psi_h \Big(|\lambda| \, \Big| \sum\limits_{t \in T} \alpha(t) Y_1(t) \Big| \Big)$

Combining (2.24) and (2.25) we have

$E \exp i \sum\limits_{t \in T} \alpha(t) V(t) = \exp - E_m \Psi_h \Big(\Big| \sum\limits_{t \in T} \alpha(t) Y_1(t) \Big| \Big).$

Thus $(V(t))_{t \in T}$ and $(X(t))_{t \in T}$ agree on cylinder sets and this is what is meant by saying that the two processes agree in distribution.

To obtain (2.22) we note that

(2.26) $\mathrm{Re}\Big(\sum\limits_{t \in T} \bar{\alpha}(t) V(t) \Big) \overset{\mathcal{D}}{=} \sum\limits_{j=1}^{\infty} F^{-1}(\Gamma_j) \mathrm{Re}\Big(k_j \tilde{g}_j \sum\limits_{t \in T} \bar{\alpha}(t) Y_j(t) \Big)$

$\overset{\mathcal{D}}{=} \sum\limits_{j=1}^{\infty} F^{-1}(\Gamma_j) h_j \Big| \sum\limits_{t \in T} \bar{\alpha}(t) Y_j(t) \Big|,$

(since for $z \in \mathbb{C}$ $\mathrm{Re}(\bar{z}\tilde{g}_1) \overset{\mathcal{D}}{=} |z| g_1$ and by definition $h = k_1 g_1$). The proof can now be completed as in the real case.

We obtain (2.23) in a similar way since in this case $\mathrm{Re}(\bar{z} e^{i\theta}) = |z| \cos(\theta - \phi)$, where ϕ depends on z. Therefore

$$\text{Re}\left(\sum_{t \in T} \overline{\alpha(t)}W(t)\right) \overset{\mathcal{D}}{=} \sum_{j=1}^{\infty} F^{-1}(\Gamma_j)\text{Re}\left(k_j e^{i\theta_j} \sum_{t \in T} \overline{\alpha(t)}Y_j(t)\right)$$

$$\overset{\mathcal{D}}{=} \sum_{j=1}^{\infty} F^{-1}(\Gamma_j)k_j \left|\sum_{t \in T} \overline{\alpha}(t)Y_j(t)\right| \cos(\theta_j - \phi_j)$$

where ϕ_j depends on Y_j. By Lemma 2.1, $\text{Re}\left(\sum_{t \in T} \overline{\alpha(t)}W(t)\right)$ is equal in distribution to a symmetric infinitely divisible random variable with Levy measure τ_η and Levy transform Ψ_η where $\eta = k_1 \left|\sum_{t \in T} \overline{\alpha(t)}Y_1(t)\right| \cos(\theta_1 - \phi_1)$. We have

$$\Psi_\eta(|\lambda|) = E_\eta \Psi(|\lambda||\eta|)$$

$$= E_m E_{k_1} \frac{1}{2\pi} \int_0^{2\pi} \Psi\left(|\lambda| \left|\sum_{t \in T} \overline{\alpha(t)}Y_1(t)\right| |[k_1 \cos(\theta_1 - \phi_1)|\right)d\theta_1$$

$$= E_m E_{k_1} \frac{1}{2\pi} \int_0^{2\pi} \Psi\left(|\lambda| \left|\sum_{t \in T} \overline{\alpha(t)}Y_1(t)\right| |k_1 \cos \theta_1|\right)d\theta_1$$

$$= E_m E_h \Psi\left(|\lambda| \left|\sum_{t \in T} \overline{\alpha(t)}Y_1(t)\right| |h|\right)$$

$$= E_m \Psi_h\left(|\lambda| \left|\sum_{t \in T} \overline{\alpha(t)}Y_1(t)\right|\right) .$$

The proof is completed as in the real case.

We will be primarily interested in (2.21) when $\{h_j\}$ is $\{\varepsilon_j\}$ or $\{g_j\}$ and in (2.22) and (2.23) when $\{k_j\}$ is $\{\varepsilon_j\}$.

As a corollary of the above results we will present a series representation of the real and complex valued infinitely divisible random variables defined in (2.1) and (2.4).

Recall that for g a mean zero normal random variable, ξ_g is, by definition a symmetric real valued infinitely divisible random variable satisfying, $\forall \lambda \varepsilon \mathbb{R}$,

$$(2.27) \qquad\qquad E e^{i\lambda \xi_g} = \exp - \Psi_g(|\lambda|)$$

where

(2.28)
$$\Psi_g(|\lambda|) = E_g \Psi(|\lambda g|)$$

and Ψ is the Levy transform corresponding to ξ (i.e. (2.1) is valid for this ξ and Ψ). We will say that a symmetric real valued infinitely divisible random variable is of type G if it is of the form ξ_g for some symmetric real valued infinitely divisible random variable ξ.

We define the class Θ as those symmetric infinitely divisible random variables which have Levy transforms of the form

(2.29)
$$\Psi_\theta(|\lambda|) = \frac{2}{\pi} \int_0^{\pi/2} \Psi(|\lambda \cos \theta|)d\theta$$

$$= \frac{2}{\pi} \int_0^1 \frac{\Psi(|\lambda u|)}{(1-u^2)^{1/2}} du \ .$$

It is clear that class Θ includes class G. The significance of class Θ is that it determines which complex valued infinitely divisible random variables can be defined as in (2.4).

Corollary 2.4: Let ξ be a symmetric real valued infinitely divisible random variable with Levy measure τ and Levy transform Ψ then

(2.30)
$$\xi \overset{\mathcal{D}}{=} \sum_{j=1}^{\infty} \epsilon_j F^{-1}(\Gamma_j)$$

where $F^{-1}(t) = \sup[u: \tau[u,\infty) > t]$.

Let Ψ_θ and Ψ be as in (2.29). For every such Ψ_θ we can define a complex valued infinitely divisible random variable $\tilde{\xi}$ satisfying $\forall z \in \mathbb{C}$

(2.31)
$$E e^{i \ \text{Re}(\overline{z}\tilde{\xi})} = e^{-\Psi_\theta(|z|)} \ ,$$

where

(2.32)
$$\tilde{\xi} \overset{\mathcal{D}}{=} \sum_{j=1}^{\infty} F^{-1}(\Gamma_j)e^{i\theta_j}$$

and $\mathrm{Re}(\tilde{\xi})$ and $\mathrm{Im}(\tilde{\xi})$ are equal in distribution and equal in distribution to $\xi_{\cos\,\theta}$ for ξ as given in (2.30).

Furthermore, the complex valued infinitely divisible random variables ξ defined in (2.4) are actually of the form (2.31), i.e. the admissible Levy transforms (2.4) are of the form of those given in (2.29).

Proof: The real case is just Lemma 2.1 with $\{h_j\}$ equal to $\{\epsilon_j\}$ since, obviously,

$$\Psi_\epsilon(|\lambda|) = E_\epsilon \Psi(|\lambda\epsilon|) = \Psi(|\lambda|) \ .$$

The complex case is a special case of (2.23) where T has one element, $\{k_j\} = \{\epsilon_j\}$ and $Y_j(t) \equiv 1$, since obviously, $h \overset{D}{=} \epsilon_1 \cos\theta_1 \overset{D}{=} \cos\theta_1$. The statements about the real and imaginary parts of $\tilde{\xi}$ follow from Lemma 2.1 and the fact that $\xi_{\cos\,\theta} \overset{D}{=} \xi_{\sin\,\theta}$.

To see that the admissible Levy transforms in (2.4) must be of the form (2.29) let $\tilde{\xi} = \xi_1 + i\xi_2$ and $z = s + it$, s, t real. Then (2.4) is

$$E \exp i(s\xi_1 + t\xi_2) = \exp - \Psi\left(|s^2 + t^2|^{1/2}\right) \ .$$

From this we see that (ξ_1,ξ_2) is a rotationally invariant infinitely divisible random variable in $\mathbb{R}^2$. Consequently there is a Levy measure μ on $\mathbb{R}^2$ such that

$$(2.33) \qquad E \exp i\left(s\xi_1 + t\xi_2\right) = \exp - \int_{\mathbb{R}^2} \left(e^{i(sx_1+tx_2)} - 1\right) d\mu(x_1,x_2) \ .$$

One can show that (ξ_1,ξ_2) is rotationally invariant on $\mathbb{R}^2$ if and only if $\mu(x_1,x_2)$ is rotationally invariant on $\mathbb{R}^2$. In this case one can write (2.33) in the form

$$\exp - \int_0^{2\pi} \int_0^\infty \left[\cos \left(\rho(s \cos \theta + t \sin \theta) \right) - 1 \right] d\tau [p,\infty) d\theta$$

$$= \exp - \int_0^{2\pi} \int_0^\infty \left[\cos(\rho \,|s^2 + t^2|^{1/2} \cos(\theta - \phi)) - 1 \right] d\tau [p,\infty) d\theta$$

$$= \exp - \int_0^{2\pi} \Psi \left(|s^2 + t^2|^{1/2} \cos(\theta - \phi) \right) d\theta$$

$$= \exp -\Psi_\theta \left(|s^2 + t^2|^{1/2} \right)$$

where ϕ is some fixed real number $\tau[p,\infty)$ is a Levy measure on $\mathbb{R}^+/\{0\}$ and Ψ and τ are related by (2.2). This completes the proof of Corollary 2.4.

Remark 2.5: Not only is the class of admissible Levy transforms in (2.4) those given by (2.29), these are also the Levy transforms for which the definition in (2.6) makes sense. To sum this up we see that for any symmetric infinitely divisible random variable ξ we can define a real-valued ξ-radial process on $\mathbb{R}^T$ by (2.5). In the complex case we can do this as long as ξ is of class Θ, i.e. those ξ with Levy transforms representable as in (2.29).

We can now describe how a process with characteristic functional given by (1.5) can be represented as a random Fourier transform. We now know that (1.5) is actually of the form

$$(2.34) \qquad E \exp i \, \mathrm{Re}\left(\sum_{j=1}^n \overline{\alpha}_j X(t_j) \right) = \exp - E_m \Psi_\theta \left(\left| \sum_{j=1}^n \overline{\alpha}_j \gamma(t_j) \right| \right)$$

where Ψ_θ is defined in terms of some Levy transform Ψ by (2.29). For latter use let us denote by τ the Levy measure associated with Ψ as in (2.1) and let F^{-1} be as defined in Corollary 2.4 for this τ. Based on Ψ_θ one can define an independently scattered infinitely divisible complex valued measure $\tilde{M}$ on (Γ, G, m) as follows: For measurable sets $A \subset \Gamma$, $B \subset \Gamma$

$$(2.35) \qquad E \exp i \, \mathrm{Re} \, (\overline{z}\tilde{M}(A)) = \exp -m(A) \, \Psi_\theta(|z|), \quad z \in \mathbb{C}$$

and, if $A \cap B = \emptyset$, $\tilde{M}(A)$ and $\tilde{M}(B)$ are independent. The complex valued stochastic process

$$(2.36) \qquad \qquad \int_\Gamma \gamma(t)\tilde{M}(d\gamma), \quad t \varepsilon K$$

has the same characteristic functional as (2.34). As in the real case it follows from [R] that if (2.34) is the characteristic functional of a probability measure on $C(K)$, the Banach space of complex valued continuous functions on K equipped with the sup-norm, or equivalently, if $\{X(t)\}_{t\varepsilon K}$ has a version with continuous sample paths then (2.36) can be defined in the usual way as a $C(K)$ valued stochastic integral.

We can give a more revealing description of $\tilde{M}(A)$. It follows from (2.2) and (2.32) that

$$(2.37) \qquad \qquad \tilde{M}(A) = \sum_{j=1}^{\infty} F^{-1}\left(\frac{\Gamma_j}{m(A)}\right) e^{i\theta_j} \ .$$

It is clear from both (2.35) and (2.37) that $\operatorname{Re} \tilde{M}(A)$ and $\operatorname{Im} \tilde{M}(A)$ are equal in distribution but are not independent. If $\operatorname{Re} \tilde{M}(A)$ is in class G the relationship between its real and imaginary parts can be even more clearly described. Using the same Ψ, τ and F^{-1} that gives rise to (2.37) we can also define

$$(2.38) \qquad \qquad \tilde{M}_1(A) = \sum_{j=1}^{\infty} F^{-1}\left(\frac{\Gamma_j}{m(A)}\right) \tilde{g}_j \ .$$

Recall that $\tilde{g}_1 = (g_1 + ig_1')$ where g_1 and g_1' are independent normal random variables with mean zero and equal variances. Let $\eta_1 = \left(g_1^2 + g_1'^2\right)^{1/2}$ and let $\{\eta_j\}_{j=1}^{\infty}$ be i.i.d. copies of η_1. Then

$$(2.39) \qquad \qquad \tilde{M}_1(A) \overset{\mathcal{D}}{=} \sum_{j=1}^{\infty} F^{-1}\left(\frac{\Gamma_j}{m(A)}\right)\eta_j e^{i\theta_j}$$

$$\overset{\mathcal{D}}{=} \sum_{j=1}^{\infty} F_\eta^{-1}\left(\frac{\Gamma_j}{m(A)}\right)e^{i\theta_j}$$

where F_η^{-1} is defined as follows: Since $g \equiv g_1 \overset{\mathcal{D}}{=} \cos \theta_1 \eta_1 \equiv \cos \theta \eta$ we have

$$\Psi_g(|\lambda|) = E_\theta E_\eta \Psi(|\lambda\eta \cos \theta|)$$

$$\equiv E_\theta \Psi_\eta(|\lambda \cos \theta|) \ .$$

Let τ_η be the Levy measure corresponding to Ψ_η. Then

$$F_\eta^{-1}(t) = \sup[u: m(A)\tau_\eta[u,\infty) > t] \ .$$

By (2.38) we see that $\operatorname{Re} \tilde{M}_1(A)$ is of class G. In (2.39) we see that $\operatorname{Re} \tilde{M}_1(A)$ is also of class Θ. Since $\operatorname{Re} \tilde{M}_1(A)$ is of class Θ, $\tilde{M}_1$ is admissible in defining a stochastic integral as in (2.36). The point of all this is that we can write

$$(2.40) \qquad \tilde{M}_1(A) \overset{D}{=} \big(\sum_{j=1}^{\infty} \big(F^{-1}(\tfrac{\Gamma_j}{m(A)})\big)^2\big)^{1/2} \ \tilde{g} = \eta_A \tilde{g}$$

where $\tilde{g} \equiv \tilde{g}_1$ and η_A^2 is a positive infinitely divisible random variable with Levy measure $m(A)\tau[\lambda^{1/2}, \infty)$. The representation in (2.40) gives us a clearer picture of the relationship between the real and imaginary parts of $\tilde{M}_1(A)$. (Of course random variables expressible as in (2.38) are only a subclass of those expressible as in (2.37))

As an example of this let $\Psi(|\lambda|) = |\lambda|^p$, $0 < p < 2$. Then $m(A)\tau[\lambda^{1/2},\infty) = m(A)\lambda^{-p/2}$ which implies that η_A is equal in distribution to $(m(A))^{1/p} \theta_{p/2}^{1/2}$, where $\theta_{p/2}$ is a positive $p/2$ stable random variable.

The next lemma further describes random variables of class G.

<u>Lemma 2.6</u>: Let ξ and ξ_g be as defined in (2.27) and (2.28). Then

$$(2.41) \qquad \xi_g \overset{D}{=} \eta g$$

where η^2 is a positive infinitely divisible random variable as defined in (2.15) and (2.16), independent of g, and with Levy measure $\tau[t^{1/2}, \infty)$ where τ is the Levy measure of ξ. Furthermore the Laplace transform of η^2,

(2.42) $E \, e^{-s\eta^2} = \exp - \Psi_g(s^{1/2}/\sigma), \quad \forall s \in R^+,$

where $\sigma\sqrt{2} = (E|g|^2)^{1/2}$. It follows from this that

(2.43) $\tilde{\xi} \overset{D}{=} \eta\tilde{g}$

where $\tilde{\xi}$ satisfies

$$E \, e^{i \, Re(\bar{z}\tilde{\xi})} = e^{-\Psi_g(|z|)} ,$$

$\tilde{g} = g + ig'$, where g' is an independent copy of g and Ψ, Ψ_g and τ are related by (2.2) and (2.28)

Proof: By Lemma 2.1 and a well known property of independent mean zero

normal random variables

(2.44) $\xi_g \overset{D}{=} \sum_{j=1}^{\infty} F^{-1}(\Gamma_j) g_j$

$$\overset{D}{=} (\sum_{j=1}^{\infty} (F^{-1}(\Gamma_j))^2)^{1/2} \, g \, .$$

By Lemma 2.2 with $h \equiv 1$

$$\sum_{j=1}^{\infty} (F^{-1}(\Gamma_j))^2$$

is a positive infinitely divisible random variable with Levy measure $\nu[t,\infty)$ which is such that $(F^{-1}(t))^2 = \sup[u: \nu[u,\infty) \leq t]$. It is easy to see that $\nu[t,\infty) = \tau[t^{1/2},\infty)$.

To obtain (2.42) we note that by (2.27) and (2.41)

$$E \exp - \Psi_g(|\lambda|) = E \, e^{i\lambda\xi_g} = E \, e^{i\lambda\eta g}$$

$$= E_\eta E_g e^{i\lambda\eta g} = E_\eta e^{-\lambda^2\sigma^2\eta^2} \, .$$

Thus (2.42) follows with $\lambda = s^{1/2}/\sigma$. Finally to obtain (2.43) we note that for $z = u + iv$

$$E \, e^{i \, Re(z\tilde{\xi})} = E_\eta E_{g,g'} \, e^{i(\eta ug + \eta ug')}$$

$$= E_\eta e^{-(u^2+v^2)\eta^2\sigma^2}$$

Therefore by (2.42) we get (2.43).

Remark 2.7: In [F] pg. 172 Feller remarks that if ξ is p-stable, $0 < p < 2$ (i.e. $\Psi(|\lambda|) = C|\lambda|^p$ for some constant C). Then $\xi = \eta g$ where η and g are independent and g is normal with mean zero. Lemma 2.6 is an extension of this observation of Feller, namely that all real valued infinitely divisible random variables of type G have such a factorization. (This factorization can also be obtained by considering Brownian motion subordinated to an infinitely divisible process with positive jumps, see e.g. XVII, 3(g), [F].) It seems reasonable to expect that if a symmetric real valued infinitely divisible random variable has such a factorization then it must be of type G but we can not prove this.

We will say that a real or complex valued ξ-radial process is of type G if the Levy transform that appears in (2.19) or (2.20) (i.e. the function $\Psi_h(\ )$ is of the form (2.28) (i.e. $h = g$ where g is a mean zero normal random variable.) ξ-radial processes of type G are marginally Gaussian. This observation was made for stable processes in [MP2] Lemma 1.6.

Lemma 2.8: (a) Let T be an index set. Let $(X(t))_{t \in T}$ be a real (resp. complex) valued ξ-radial process of type G. Then we can find probability spaces (Ω, F, P) and (Ω', F', P') and a real (resp. complex) valued stochastic process $(\chi(t))_{t \in T}$ defined on $(\Omega, F, P) \times (\Omega', F', P')$ such that

(i) The processes $(\chi(t))_{t \in T}$ and $(X(t))_{t \in T}$ have the same distribution.

(ii)   For each fixed $\omega \varepsilon \Omega$, the random process $(\chi(t;\omega,\cdot))_{t\varepsilon T}$ is a
real valued Gaussian process, (resp. complex vlaued Gaussian
process with independent real and imaginary parts.)

(iii) If $(X(t))_{t\varepsilon T}$ has a version with continuous sample paths than
for almost all $\omega \varepsilon \Omega$, the random process $(\chi(t;\omega,\cdot))_{t\varepsilon T}$ is
a continuous real-valued Gaussian process, (resp. continuous
complex valued Gaussian process with independent real and
imaginary parts.)

(b) Moreover, if T is a locally compact Abelian group G and if
$(X(t))_{t\varepsilon T}$ is strongly stationary, then we can find $(X(t))_{t\varepsilon T}$ as above
verifying the additional condition: For each fixed $\omega \varepsilon \Omega$, the process
$(\chi(t;\omega,\cdot))_{t\varepsilon T}$ is a stationary real valued Gaussian process, (resp.
stationary complex valued Gaussian process with independent real and
imaginary parts.)

Proof: The proof follows from Lemma 2.3. Let $\{\Gamma_j\}$, $\{Y_j\}$, $\{h_j\}$, $\{g_j\}$,
$\{\tilde{g}_j\}$ and $\{k_j\}$ be as in Lemma 2.3. To prove (i) and (ii) of part (a)
in the real case we use (2.21) with $\{h_j\} = \{g_j\}$ defined on (Ω',A',P')
and $\{\Gamma_j\}$ and $\{Y_j\}$ defined on (Ω,A,P). For (i) and (ii) of (a) in
the complex case we use (2.22) and set $k_j \equiv 1$ and let $\{\tilde{g}_j\}$ be
defined on (Ω',A',P') and $\{\Gamma_j\}$ and $\{Y_j\}$ be defined on (Ω,A,P).
If $\{F^{-1}(\Gamma_j)\}_{j=1}^{\infty}$ were independent then (iii) of part (a) would be an
immediate consequence of the Ito-Nisio theorem. Nevertheless similar
arguments show that, since (2.21) and (2.22) are marginally independent,
if $(X(t))_{t\varepsilon T}$ has a version with continuous paths in the real (resp.
complex) case then the series (2.21) (resp. (2.22)) converges uniformly
a.s. (For further details see Remark 4.4, Chapter II, [MP1].)

We will use the following notation with respect to metric entropy.
Let (T,d) be a complex metric space equipped with a pseudo-metric d.

We will denote by $N(T,d;\varepsilon)$ the smallest number of open balls in the pseudo-metric d which covers T. We introduce the function $\sigma(T,d;n)$ which is defined, for each integer n by

$$(2.45) \qquad \sigma(T,d;n) = \inf\{\varepsilon > 0: \ N(T,d;\varepsilon) \leq n\}.$$

Let η and H_η be as defined in (1.8) and (1.9). It is not difficult to see that

$$(2.46) \qquad \int_0^\infty H_\eta(\log N(T,d;\varepsilon))d\varepsilon < \infty \quad \text{iff} \quad \sum_{n=3}^\infty \frac{\sigma(T,d;n)}{n\eta^{-1}(\log n)} < \infty \ .$$

Indeed, we know that the left side of (2.46) is finite iff

$$(2.47) \qquad \int_0^\infty \varepsilon \ dH_\eta(\log N(T,d;\varepsilon)) < \infty$$

(see e.g. Lemma 3.6, Chapter II, [MP1]). Furthermore the integral in (2.47)

$$= \sum_{n=2}^\infty \sigma(T,d;n) \int_{(\log n)\vee 1}^{\log(n+1)} \frac{ds}{\eta^{-1}(s)} \ .$$

It is clear that this last sum converges iff the right side of (2.46) converges. Furthermore if $\eta(x)$ is regularly varying at infinity of index $\rho > 1$ then

$$(2.48) \qquad \int_1^x \frac{ds}{\eta^{-1}(x)} \sim \frac{x}{\eta^{-1}(x)} \ , \ x \geq 1 \ .$$

(Here we use the notation $f(x) \sim g(x)$ to mean there exist constants $c_1 > 0$, c_2 such that $c_1 g(x) \leq f(x) \leq c_2 g(x)$.) Finally let us observe that by a change of variables and the monotonicity of $\sigma(T,d;n)$ it is easy to see that if

$$(2.49) \qquad \sum_{n=n_0(\alpha)}^\infty \frac{\sigma(T,d;n)}{n\eta^{-1}(\alpha \log n)} < \infty$$

for some $0 < \alpha < \infty$ then it is finite for all $0 < \alpha < \infty$, where $n_0(\alpha)$ is simply large enough so that $\eta^{-1}(\alpha \log n)$ is properly defined.

In a completely analogous fashion we see that

(2.50) $\int_0^\infty \overline{H}_\nu (\log N(K,d_{Y,\phi};\epsilon))d\epsilon < \infty$

                iff $\sum_{n=3}^\infty n^{-1}\sigma(K,d_{Y,\phi};n)\nu^{-1}(\frac{1}{\alpha \log n}) < \infty$

for some $0 < \alpha < \infty$, where ν and $\overline{H}_\nu$ are defined in (1.22) and
(1.23).

We end this section with an explanation of why the processes
defined in (2.5) and (2.6) are called ξ-radial. It follows from [AG]
that a symmetric infinitely divisible probability measure without a
Gaussian component on a Banach space B has a unique characteristic
functional given by

(2.51) $\exp\left[\int_B (\cos f(x) - 1)\, d\mu(x)\right]$, $f \in B^*$

where μ is a symmetric measure on B. Suppose that μ factors in
such a way that for A a measurable symmetric set on the unit sphere S
of B and for D a measurable set in $\mathbf{R}^+$ we have

                $\mu\{x \in B : \frac{x}{\|x\|} \in A,\ \|x\| \in D\} = m(A)\tau(D)$

where m is a probability measure and τ is a Levy measure, i.e. τ
satisfies (2.3). Then (2.51)

(2.52) $= \exp -\int_S \int_0^\infty (\cos f(s\|x\|) - 1))\, d\tau[\|x\|, \infty)\, m(ds)$

                $= \exp -\int_S \Psi(|f(s)|)\, m(ds)$

where Ψ is given in (2.2). This is what we have in (2.5) and (2.6).
It is no longer necessary to stipulate that m is symmetric on S
since the expression in (2.52) "symmetrizes" m. Therefore the measure
μ, which is called a Levy measure, has identically distributed radial
components. The term ξ-radial identifies τ by the associated real

valued symmetric infinitely divisible random variable ξ as defined in
(2.1) and (2.2). As is well known a p-stable measure, $0 < p < 2$, on a
Banach space is ξ-radial with $\tau[\lambda,\infty) = \lambda^{-p}$, $\lambda > 0$ and with ξ a real
valued symmetric p-stable random variable. Finally note that although
we have just discussed infinitely divisible probability laws on a Banach
space the problem is to figure out when such laws exist. Thus we begin
with cylinder set measures in (2.5) and (2.6) and then take up the
question of whether they determine measures on the space of continuous
functions.

3. NECESSARY CONDITIONS FOR CONTINUITY

Let (Ω, A, P) and (Ω', A', P') be probability spaces and $(\chi(t))_{t \in T}$ be a real or complex valued ξ-radial process of type G defined on $(\Omega, A, P) \times (\Omega', A'P')$ satisfying (ii) in Lemma 2.8. This means that $\chi(t)$ has characteristic functional given by (2.19) in the real case and (2.20) in the complex case with $h = g$ a mean zero normal random variable.

For each $\omega \in \Omega$ we consider the marginal Gaussian processes $((\chi(t;\omega,\cdot))_{t \in T}$. For these processes we define the random pseudo-metrics d_ω on T in the real (resp. complex) case by

$$d_\omega(s,t) = \left(\frac{1}{2} E_{\omega'} |\chi(t) - \chi(s)|^2\right)^{1/2}$$

$$\left(\text{resp. } d_\omega(s,t) = \frac{1}{2}\left(E_{\omega'} |\chi(t) - \chi(s)|^2\right)^{1/2}\right).$$

We need to slightly expand the definition of $d_{\chi,\Phi}$ given in (1.6). Let Φ be given as in (1.6), $\forall s, t \in T$ let

(3.0) $$d_{\chi,\Phi}(s,t) = \inf\left\{c > 0: \ E_m \Phi\left(\frac{|\beta(t) - \beta(s)|}{c}\right) \leq 1\right\}$$

where m and β are given in (2.19) and (2.20). The next lemma is the main result of this section.

Lemma 3.1: Let $(\chi(t))_{t \in t}$ be as above and assume that the associated Levy transform Ψ_g satisfies

(3.1) $$\Psi_g(x) \geq k_1' T(x), \quad \forall x \geq x_0' \geq 0$$

for some constant $k_1' > 0$ and that

(3.2) $$T(xy) \geq k_2' \eta(x)\Phi(y), \quad \forall y \geq y_1' \geq 0, \quad x \geq x_1' \geq 1$$

for some constant $k_2' > 0$, where $\eta(x)$ is strictly increasing for $x \geq x_1'$, $\eta(1) \geq 1$ and satisfies $\lim_{x \to \infty} \eta(x) = \infty$ and where $\Phi: R^+ \to R^+$, $\Phi(0) = 0$, is a convex function. Then for every $\varepsilon > 0$ and $\alpha < \infty$ sufficiently large there exists a subset $\Omega_0 \subset \Omega$ with $P(\Omega_0) > 1-\varepsilon$ and an integer n_0 such that for all integers $n > n_0$ and each $\omega \varepsilon \Omega_0$

$$(3.3) \qquad \sigma(T,d_\omega;n) \geq C \frac{\sigma(T,d_{\chi,\Phi};n)(\log n)^{1/2}}{\eta^{-1}(\alpha \log n)}$$

where $\sigma(T,d_\omega;n)$ and $\sigma(T,d_{\chi,\Phi};n)$ are as defined in (2.31) and $C > 0$ is a constant independent of the measure m.

The proof of Lemma 3.1 is based on the following estimate.

Lemma 3.2: With the notation of Lemma 3.1 we have that for any $\delta > 0$ there exists an $n \geq n_0(\delta)$ sufficiently large such that

$$(3.4) \qquad P\left[\omega \varepsilon \Omega: d_\omega(s,t) < \frac{(\log n)^{1/2}}{\eta^{-1}(\alpha \log n)} d_{\chi,\Phi}(s,t)\right] \leq n^{-\delta} \quad \forall s, t \varepsilon T,$$

where α is a constant, depending on δ, but finite for δ finite, and independent of m.

Proof: If $d_{\chi,\Phi}(s,t) \equiv 0$, $\forall s, t \varepsilon T$ there is nothing to prove. Thus we rule out this case. We will give the proof for $(\chi(t))_{t \varepsilon T}$ complex. The proof in the real case is completely similar. Since $\chi(t;\omega,\cdot)$ is a Gaussian process on Ω', we have, with $z = 2^{-1/2}(1+i)\lambda$, λ real, that

$$E_{\omega'} \exp i \ \mathrm{Re}[\overline{z}(\chi(t;\omega,\omega') - \chi(s;\omega,\omega'))] \ = \ \exp -|\lambda|^2 d_\omega^2(s,t) \ .$$

On the other hand, by (2.20) with the same value of z we get

$$E_\omega E_{\omega'} \exp i \ \mathrm{Re}[\overline{z}(\chi(t;\omega,\omega') - \chi(s,\omega,\omega'))] = \exp - E_m \Psi_g(|\lambda||\beta(t) - \beta(s)|) \ .$$

Combining these two equations we have

(3.4a) $E_\omega \exp - |\lambda|^2 d_\omega(s,t) = \exp - E_m \Psi_g(|\lambda||\beta(t) - \beta(s)|).$

Therefore, by (3.4a) and an exponential Chebysev inequality we get,
$\forall \varepsilon > 0,\ \lambda$ real

(3.5) $P\{\omega\ \varepsilon\ \Omega:\ d_\omega(s,t) < \varepsilon d_{X,\Phi}(s,t)\}$

$$\leq (E_\omega \exp -|\lambda|^2 d_\omega^2(s,t))\ \exp\ \varepsilon^2 \lambda^2 d_{X,\Phi}^2(s,t)$$

$$= \exp\left[-E_m \Psi_g(|\lambda||\beta(s) - \beta(t)|) + \varepsilon^2 \lambda^2 d_{X,\Phi}^2(s,t)\right].$$

Since Φ is convex we can find a convex function $\tilde{\Phi}$ such that for
some fixed $\tau > 0$

(3.6) $\tilde{\Phi}(y) = 0,$ $0 \leq y \leq y_1'$

(3.7) $\tilde{\Phi}(y) \geq \Phi(y),$ $\forall y \geq y_0' \geq y_1'$

and

(3.8) $\tilde{\Phi}(y) \leq (1+\tau)\Phi(y),\quad \forall y \geq 0$.

Note that by (3.1) and (3.2)

(3.9) $\Psi_g(xy) \geq k_1' k_2' \eta(x)\Phi(y),\quad \forall y \geq y_1',\quad x \geq x_1' \vee x_0'/y_1' \equiv x_3.$

Therefore, by (3.6) and (3.8)

(3.14) $\Psi_g(xy) \geq C\eta(x)\tilde{\Phi}(y),\quad \forall y \geq 0,\quad x \geq x_3,$

where $C = k_1' k_2'/(1+\tau).$

Consider $d_{X,\tilde{\Phi}}(s,t)$ as defined in (3.0) for this convex function
$\tilde{\Phi}$. Because of (3.7) there exists a constant C_3 such that

(3.15) $d_{X,\Phi}(s,t) \leq C_3 d_{X,\tilde{\Phi}}(s,t),\quad \forall s,\ t\ \varepsilon\ T$.

This is a well known fact, see e.g. [KR]). By (3.5) with $\lambda = \rho/d_{X,\tilde{\Phi}}(s,t)$
for $\rho > x_3,$ we have

(3.16) $P\{\omega \in \Omega: d_\omega(s,t) \leq \varepsilon d_\Phi(s,t)\}$

$$\leq \exp\left[-E_m \Psi_g\left(\frac{|\beta(s) - \beta(t)|}{d_{X,\tilde{\Phi}}(s,t)}\rho\right) + \varepsilon^2\rho^2 \frac{d^2_{X,\Phi}(s,t)}{d^2_{X,\tilde{\Phi}}(s,t)}\right]$$

which by (3.14) and (3.15)

(3.17) $$\leq \exp\left[-C_2 E_m \tilde{\Phi}\left(\frac{|\beta(s) - \beta(t)|}{d_{X,\tilde{\Phi}}(s,t)}\right) \eta(\rho) + \varepsilon^2\rho^2 C_3^2\right].$$

By the definition of $d_{X,\tilde{\Phi}}$ and the continuity of $\tilde{\Phi}$ the last term is

(3.18) $$= \exp\left[-C_2\eta(\rho) + \varepsilon^2\rho^2 C_3^2\right].$$

Choose $\alpha = (\delta + C_3^2)/C_2$ and n_0 such that $\alpha \log n_0 > x_1$ and $n^{-1}(\alpha \log n_0) > x_3$. For $n \geq n_0$ set $\rho = n^{-1}(\alpha \log n)$ and $\varepsilon^2 = \log n/\rho^2$. Then for $n \geq n_0$, (3.17)

(3.19) $$= \exp\left[-C_2 \alpha \log n + C_3^2 \log n\right] = n^{-\delta}.$$

Combining (3.16) - (3.19) we get (3.4)

Remark 3.3: The reader might wonder why $\tilde{\Phi}$ was introduced. This is because our primary objective at this point is to prove Theorem 1.1 part II.) In (1.10) we only require that $\Psi(|x|) \geq k_1'\Phi(x)$, $\forall x \geq x_0'$ where x_0' may be arbitrarily large. This is because boundedness (or unboundedness) of ξ-radial processes is only determined by $\Psi(|x|)$ for $|x|$ large. Substituting $\tilde{\Phi}$ for Φ enables us to ignore the values of $\Phi(x)$ for x small.

Proof of Lemma 3.1: Lemma 3.2 is the basic inequality. One can now follow "proof of Lemma 2.1" in [MP2] practically word for word, and arrive at the inequality

(3.20) $P\left[\bigcup_{n=n_0}^{\infty} \left\{\omega: \sigma(T,d_\omega;n) < \dfrac{(\log n)^{1/2}}{2n^{-1}(\alpha \log n)} \sigma(T,d_{\chi,\phi};n)\right\}\right]$

$$\leq \sum_{n=n_0}^{\infty} n^{2-\delta} \ .$$

It is clear that one can find a δ and n_0 sufficiently large so that the right side of (3.20) is less than ε. This completes the proof of Lemma 3.1 since we can take

$$\Omega_0 = \bigcap_{n=n_0}^{\infty} \left\{\omega: \sigma(T,d_\omega;n) \geq \dfrac{(\log n)^{1/2}}{2n^{-1}(\alpha \log n)} \sigma(T,d_{\chi,\phi};n)\right\} \ .$$

We can now prove the necessary part of Theorem 1.1.

<u>Proof of Theorem 1.1, part I.)</u>: The beginning of this proof is essentially verbatim the proof of Theorem 2.9, [MP2]. Consider $\{X(t)\}_{t\varepsilon K}$ as given in the hypothesis of this theorem but with the additional condition that it is a process of type G. Assume that $\{X(t)\}_{t\varepsilon K}$ has a version with continuous sample paths. By Lemma 2.7 we can define $\{X(t)\}_{t\varepsilon K}$ on a product space $(\Omega,A,P) \times (\Omega',A',P')$ such that for each fixed $\omega \varepsilon \overline{\Omega}$, $\overline{\Omega} \subset \Omega$, $P(\overline{\Omega}) = 1$, the process $\{X(t;\omega,\cdot)\}_{t\varepsilon K}$ is a continuous stationary Gaussian process. Thus for each $\omega \varepsilon \overline{\Omega}$, we have by Fernique's theorem that

(3.21) $\sum_{n=1}^{\infty} \dfrac{\sigma(K,d_\omega;n)}{n(\log(n+1))^{1/2}} < \infty \ .$

By Lemma 3.1 there is an n_0 and a subset $\overline{\Omega}' \subset \overline{\Omega}$ with $P(\overline{\Omega}) > 1/2$ such that for $n \geq n_0$

$$\sum_{n=n_0}^{\infty} \dfrac{\sigma(K,d_\omega;n)}{n(\log(n+1))^{1/2}} \geq c' \sum_{n=n_0}^{\infty} \dfrac{\sigma(K,d_{\chi,\phi};n)}{nn^{-1}(\alpha \log n)}$$

for all $\omega \in \overline{\Omega}'$, where $C' > 0$ is a constant independent of m. By (3.21) this implies that

$$\sum_{n=3}^{\infty} \frac{\sigma(K,d_{X,\Phi};n)}{nn^{-1}(\alpha \log n)} < \infty$$

or, equivalently, by (2.46) and (2.49) that

(3.22) $\int_0^{\infty} H_\eta(\log N(K,d_{X,\Phi};\epsilon))d\epsilon < \infty$.

Thus if $\{X(t)\}_{t\in T}$ has a version with continuous paths (3.22) holds. This implies (1.12) for processes of type G.

We now remove the condition that the process must be of type G. Let us consider the general case of a real valued process with characteristic functional given by (1.3) where the Levy transform Ψ satisfies (1.10). By (2.21) of Lemma 2.3, this process can be represented by the series

(3.23) $\sum_{j=1}^{\infty} F^{-1}(\Gamma_j)\epsilon_j Y_j(t), \quad t \in K$

where $\{\epsilon_j\}$ is a Rademacher sequence and F^{-1} is as defined in (2.18) where τ is the Lévy measure corresponding to Ψ. Recall that the sequences $\{\Gamma_j\}_{j=1}^{\infty}$, $\{\epsilon_j\}_{j=1}^{\infty}$ and $\{Y_j\}_{j=1}^{\infty}$ are all independent. Thus we can define the series in (3.23) on the product probability space $(\Omega_1, F_1, P_1) \times (\Omega_2, F_2, P_2) \times (\Omega_3, F_3, P_3)$ and write it as

(3.24) $\sum_{j=1}^{\infty} F^{-1}(\Gamma_j(\omega_1))\epsilon_j(\omega_2)Y_j(t,\omega_3)$.

Fix ω_1 and ω_3 and consider the related series

(3.25) $\sum_{j=1}^{\infty} F^{-1}(\Gamma_j(\omega_1)g_j(\omega_2)Y_j(t,\omega_3)$.

Both of these series are random Fourier series on (Ω_2, F_2, P_2). It follows

from Theorem 1.1 Chapter I, [MP1] that for ω_1 and ω_3 fixed either both
the series (i.e. (3.24) and (3.25)) converge uniformly a.s. with respect
to (Ω_2, F_2, P_2) or else they both represent processes with unbounded sample
paths. It follows then, by Fubini's theorem, that if the process in
(3.23) has a version with continuous sample paths then so does the process
of type G represented by

$$(3.26) \qquad \sum_{j=1}^{\infty} F^{-1}(\Gamma_j) g_j Y_j(t), \quad t \in K \ .$$

We now complete the proof of this theorem in the real case by showing that
under (1.12) the process in (3.26) has unbounded sample paths a.s. This
process also has a characteristic functional of the form (1.3) but with Ψ
replaced by Ψ_g where $\Psi_g(|x|) = E_g \Psi(|gx|)$. We have that

$$(3.27) \qquad \Psi_g(|x|) \geq C k_1' T(x) \qquad \forall x \geq x_0'$$

where k_1' and $\Phi(x)$ are as given in (1.10) and $C = \text{Prob}(|g| \geq 1)$. This
is easy to see since by (1.10) for $x \geq x_0'$ and $|g| \geq 1$

$$\Psi(|gx|) \geq k_1' T(x).$$

The process in (3.26) is of type G. Given (3.27) and (1.11) we know from
the earlier portion of this proof that (1.12) implies that the process in
(3.26) has sample paths that are unbounded a.s. This completes the proof
of Theorem 1.1 part I.) in the real case.

It follows from Corollary 2.4 that every complex valued ξ-radial
process can be represented by the series

$$(3.28) \qquad \sum_{j=1}^{\infty} F^{-1}(\Gamma_j) e^{i\theta_j} Y_j(t), \quad t \in K$$

where F^{-1} is defined in (2.18). Just as in the real case the process in
(3.28) has a version with continuous paths if and only if the
corresponding process of type G

$$(3.29) \qquad \sum_{j=1}^{\infty} F^{-1}(\Gamma_j)\tilde{g}_j Y_j(t), \quad t \in K$$

does. Now suppose the Levy transform (in (1.5)) that determines the process in (3.28) satisfies (1.10), then so does the Levy transform that determines the process in (3.29). This is obvious since $\Psi(|gx|) \geq \Psi(|x \cos \theta|)$ on the set where $|g| \geq 1$. Exactly as in the real case, given (1.10), (1.11) and (1.12) the process in (3.29) has unbounded sample paths a.s. and therefore so does the process in (3.28). This completes the proof of Theorem 1.1 part I.)

Consider the real valued random variable ξ defined in (1.17). Let g be a normal random variable with mean zero and variance 1. We define ψ_g by the following relationship

$$(3.30) \qquad E \, e^{i\lambda \xi g} = E_g \, e^{-\psi(|\lambda||g|)} = e^{-\psi_g(|\lambda|)} .$$

We will need the next lemma in the proof of Theorem 1.2, part I).

Lemma 3.4: Let ψ and ψ_g be related by (1.17) and (3.30) and suppose that ψ satisfies (1.24). Then there exists an $x_0' > 0$ and $k_1'' > 0$ such that

$$(3.31) \qquad \psi_g(|x|) \geq k_1'' \phi(x) , \qquad 0 \leq x \leq x_0' .$$

Proof: Clearly by the continuity of $\psi(|\lambda|)$ and the fact that $\psi(0) = 0$ we can find an a such that

$$(3.32) \qquad 1 - e^{-\psi(|\lambda|)} \geq \tfrac{1}{2} \psi(|\lambda|) , \qquad 0 \leq |\lambda| \leq a .$$

For convenience let us take $a \leq x_0$, where x_0 is given in (1.24). We have

$$(3.33) \qquad \psi_g(|\lambda|) \geq E_g[1 - e^{-\psi(|\lambda||g|)}] .$$

Now let Ω be the set on which $1 \leq |g| \leq 2$. On this set we have by (3.32) and (1.24) that for $0 < |\lambda| \leq \alpha/2$

(3.34)
$$1 - e^{-\psi(|\lambda||g|)} \geq \frac{1}{2} \psi(|g\lambda|) \geq \frac{1}{2} k_1' \phi(|g\lambda|)$$

$$\geq \frac{1}{2} k_1' \phi(|\lambda|)$$

where we also use the monotonicity of ϕ. Using (3.34) in (3.33) we get (3.31) with $k" = \frac{1}{2} k' \, P(1 \leq |g| \leq 2)$.

Proof of Theorem 1.2 part I.): Let g and g' be independent normal random variables with mean zero and variance 1. Let $\{g_\gamma\}_{\gamma \in A}$ be i.i.d. copies of g indexed by the elements of $A \subset \Gamma$ as in (1.17) and above. Similarly let $\{\tilde{g}_\gamma\}_{\gamma \in A}$ be i.i.d. copies of the complex valued normal random variable $g + ig'$. Furthermore both of these sequences are taken to be independent of $\{\xi_\gamma\}_{\gamma \in A}$. It follows from Theorem 1.1, Chapter 1, [MP1] that the uniform convergence or unboundedness a.s. of the following three series are equivalent:

(3.35)
$$\sum_{\gamma \in A} a_\gamma \xi_\gamma \gamma(t), \qquad t \in K,$$

(3.36)
$$\sum_{\gamma \in A} a_\gamma \xi_\gamma g_\gamma \gamma(t), \qquad t \in K$$

and

(3.37)
$$\sum_{\gamma \in A} a_\gamma \xi_\gamma \tilde{g}_\gamma \gamma(t), \qquad t \in K.$$

Our method of proof is the following: if $\{a_\gamma\}_{\gamma \in A}$ is a real (complex) sequence we show that (3.35) is unbounded a.s. by showing that (3.36), ((3.37)) is unbounded a.s. Actually we will just give the proof in the complex case. The real case is completely similar. The proof is parallel to the proof of Theorem 1.1 part I.) The only differences are that the functions are defined near zero rather than near infinity and that the metric is different. Let us define $\{\xi_\gamma\}_{\gamma \in A}$ on a probability space

(Ω, F, P) and $\{\tilde{g}_\gamma\}_{\gamma \in A}$ on a probability space (Ω', F', P'). Let $\omega \in \Omega$ and $\omega' \in \Omega'$. Let us denote the series in (3.37) by $\{y(t), t \in K\}$ and consider it as defined on the probability space $(\Omega \times \Omega', F \times F', P \times P')$. Clearly for each $\omega \in \Omega$

(3.38) $$y(t;\omega,.) = \sum_{\gamma \in A} a_\gamma \xi_\gamma(\omega) \tilde{g}_\gamma \gamma(t), \quad t \in K$$

is a complex valued stationary Gaussian process. As we did in the beginning of this section we associate with this process the random pseudo-metric

(3.39) $$d_\omega(s,t) = \frac{1}{2} E_{\omega'}\left(\left|y(t;\omega,\omega') - y(s;\omega,\omega')\right|^2\right)^{1/2}.$$

Now consider $d_{Y,\phi}(s,t)$ as defined in (1.21) which is associated with (1.18) or, since they are the same, (3.35). Analogous to Lemma 3.2 we first show that for any $\delta > 0$ there exists an $n \geq n_0(\delta)$ sufficiently large such that

(3.40) $$P\left[\omega \in \Omega: \ d_\omega(s,t) < (\log n)^{1/2} \nu^{-1}\left(\frac{1}{\alpha \log n}\right) d_{Y,\phi}(s,t)\right]$$

$$\leq n^{-\delta}, \quad \forall s,t \in K,$$

where α is a constant, depending on δ, but finite for δ finite and independent of $\{a_\gamma\}_{\gamma \in A}$. To obtain (3.40) we repeat the argument of the proof of Lemma 3.2. Let $z = 2^{-1/2}(1+i)\lambda$ where λ is real. We have

(3.41) $$E_{\omega'} \exp i \ \text{Re}\left[\bar{z}\left(y(t;\omega,\omega') - y(s;\omega,\omega')\right)\right] = \exp - |\lambda|^2 d_\omega^2(s,t).$$

On the other hand, with the same value of z, and taking (1.17) and (3.30) into account we have

(3.42) $$E \exp i \ \text{Re}\left[\bar{z}(y(t) - y(s))\right] = E_{\omega'} \exp - \sum_{\gamma \in A} \psi\left(|\lambda| |a_\gamma| |\gamma(s)-\gamma(t)| |g|\right)$$

$$= \exp - \sum_{\gamma \in A} \psi_g\left(|\lambda| |a_\gamma| |\gamma(s)-\gamma(t)|\right).$$

Using (3.41) and (3.42) and an exponential Chebyshev inequality we have

(3.43) $P\left(\omega \; \epsilon \; \Omega \colon \; d_\omega(s,t) < \epsilon \; d_{Y,\phi}(s,t)\right)$

$$\leq \exp\left[-\sum_{\gamma \epsilon A} \psi_g\left(|\lambda||a_\gamma||\gamma(s)-\gamma(t)|\right) + \epsilon^2\lambda^2 \; d^2_{Y,\phi}(s,t)\right].$$

We know from (3.31) that $\psi_g(|x|) \geq k''_1\phi(x)$, $0 \leq x \leq x'_0$ but let us assume for the time being that

(3.44) $\psi_g(|x|) \geq k''_1\phi(x)$, $\quad \forall x \geq 0$.

Given (3.44) the last term in (3.43)

(3.45) $\leq \exp\left[-k''_1 \sum_{\gamma \epsilon A} \phi\left(|\lambda||a_\gamma||\gamma(s)-\gamma(t)|\right) + \epsilon^2\lambda^2 d^2_{Y,\phi}(s,t)\right]$.

Let

(3.46)
$$\lambda = \left[d_{Y,\phi}(s,t)\nu^{-1}\left(\frac{1}{\alpha \log n}\right)\right]^{-1}$$

$$\epsilon = (\log n)^{1/2} \nu^{-1}\left(\frac{1}{\alpha \log n}\right)$$

and note that by (1.25)

$$\phi\left(\frac{|a_\gamma||\gamma(s)-\gamma(t)|}{d_{Y,\phi}(s,t)\nu^{-1}\left(\frac{1}{\alpha \log n}\right)}\right) \geq \frac{\alpha}{k'_2} \phi\left(\frac{|a_\gamma||\gamma(s)-\gamma(t)|}{d_{Y,\phi}(s,t)}\right) \log n$$

so that (3.45)

(3.47) $\leq \exp\left[-\dfrac{ak''_1}{k'_2} \log n + \log n\right]$

where, of course, we use the fact that

$$\sum_{\gamma \epsilon A} \phi\left(\frac{|a_\gamma||\gamma(s)-\gamma(t)|}{d_{Y,\phi}(s,t)}\right) = 1 \; .$$

Using (3.43) and (3.45) – (3.47) we get (3.40). Then following the proof that Lemma 3.2 implies Lemma 3.1 we have that for every $\epsilon > 0$ and $\alpha < \infty$

sufficiently large there exists a subset $\Omega_0 \subset \Omega$ with $P(\Omega_0) > 1-\varepsilon$ and an integer n_0 such that for all integers $n \geq n_0$ and each $\omega \in \Omega_0$

$$(3.48) \qquad \sigma(K,d_\omega;n) \geq C\sigma(K,d_{Y,\phi};n)\nu^{-1}\Big(\frac{1}{\alpha \log n}\Big)(\log n)^{1/2}$$

where $\sigma(K,d_\omega;n)$ and $\sigma(K,d_{Y,\phi};n)$ are defined in (2.45) and $C > 0$ is a constant. Therefore, following the proof of Theorem 1.1 part I.) we see that if the series in (3.37) has a version with continuous sample paths then

$$(3.49) \qquad \sum_{n=3}^{\infty} n^{-1}\sigma(K,d_{Y,\phi};n)\nu^{-1}\Big(\frac{1}{\alpha \log n}\Big) < \infty \ .$$

Using (2.50) we complete the proof of Theorem 1.2 part I.) as long as (3.44) holds.

We now remove the restriction that (3.44) holds. Recall that we are studying the process in (3.35) and the proof is being given in the case where $a_\gamma\gamma$ is complex. Instead of considering the series in (3.37) as we did in the first part of this proof let us consider

$$(3.50) \qquad Z(t) = \sum_{\gamma \in A} a_\gamma\big(\xi_\gamma\tilde{g}_\gamma + \tilde{n}_\gamma\big)\gamma(t), \quad t \in K$$

where $\{\tilde{n}_\gamma\}_{\gamma \in A}$ is an independent copy of $\{\tilde{g}_\gamma\}_{\gamma \in A}$ which is also independent of $\{\xi_\gamma\}_{\gamma \in A}$. We define $\{\tilde{\xi}_\gamma\}_{\gamma \in A}$ on (Ω, F, P) and $\{\tilde{g}_\gamma\}_{\gamma \in A}$ and $\{\tilde{n}_\gamma\}_{\gamma \in A}$ on (Ω', F', P'). Following the first part of this proof we fix $\omega \in \Omega$ and consider the marginal Gaussian process

$$Z(t;\omega,\cdot) = \sum_{\gamma \in A} a_\gamma\big(\xi_\gamma(\omega)\tilde{g}_\gamma + \tilde{n}_\gamma\big)\gamma(t), \quad t \in K \ .$$

Analogous to (3.41) we have for $z = 2^{-1/2}(1 +i)\lambda$ with λ real, that

$$E_{\omega'} \exp i \ \mathrm{Re}\big(\overline{z}\big(Z(t;\omega,\omega') - Z(s;\omega,\omega')\big)\big) = \exp - \lambda^2 d_\omega^2(s,t)$$

and

$$E \exp \operatorname{Re}\left(\overline{z}\left(Z(t) - Z(s)\right)\right) = \exp - \sum_{\gamma \in A} \psi_g\left(|\lambda| \, |a_\gamma| \, |\gamma(s) - \gamma(t)|\right)$$

$$- \frac{1}{2} |\lambda|^2 \sum_{\gamma \in A} |a_\gamma|^2 \, |\gamma(s) - \gamma(t)|^2$$

$$\equiv \exp -\tilde{\psi}\left(|\lambda| \, |a_\gamma| \, |\gamma(s) - \gamma(t)|\right) \, ,$$

where

$$\tilde{\psi}(|x|) = \psi_g(|x|) + \frac{1}{2} x^2 \, .$$

We have already seen in Lemma 3.4 that there exists an $x_0' > 0$ such that

(3.51)
$$\psi_g(|x|) \geq k_1'' \phi(x), \quad 0 \leq x \leq x_0' \, .$$

Furthermore by (1.19)

(3.52)
$$\frac{1}{2} x^2 \geq \frac{1}{2c_2} \phi(x) \qquad \text{for} \quad x \geq x_1'$$

for some x_1' sufficiently large. However we also have that for $x_0' \leq x \leq x_1'$

$$\frac{1}{2} x^2 \geq \frac{1}{2} (x_0')^2 \geq \frac{1}{2}(x_0')^2 \frac{\phi(|x|)}{\phi(|x_1'|)}$$

since ϕ is increasing. Therefore we see that there exists a constant $C > 0$ such that

$$\tilde{\psi}(|x|) \geq C\phi(x) \, , \, \forall x \geq 0 \, .$$

Thus $\tilde{\psi}$ satisfies (3.44) and so by the first part of this proof which was valid in this case we see that under (1.24), (1.25) and (1.26) the series in (3.50) has unbounded sample paths a.s. Finally we note that it follows from Theorem 1.1, Chapter I [MP1] that the series (3.50) and those in (3.35) - (3.37) are all uniformly convergent a.s. or else are unbounded a.s. This concludes the proof of Theorem 1.2 part I.).

Remark 3.5: It is probably useful to explain some of the conditions imposed on ψ and ϕ in Theorem 1.2 part I.) In (1.24) we only require that ψ is bounded below near zero. This is appropriate since the tail of the probability distribution of ξ depends on its characteristic function near the origin. We need not be concerned if ξ is bounded, that case is handled in Theorem 1.1, Chapter I [MP1] as long as $\{a_\gamma\}_{\gamma \in A} \in \ell_2$. This is assumed by the lower bound for $\phi(x)$ in (1.19). It seems strange that we also need an upper bound on ϕ for x large since $d_{\gamma,\phi}$ only depends on $\phi(x)$ for $x \leq 1$. However this upper bound implies that $\nu(x) \geq$ Const. x^2 for x near zero, which implies that

$$\nu^{-1}\left(\frac{1}{\alpha \log n}\right) \geq \frac{1}{\alpha(\log n)^{1/2}}$$
and this is necessary if our results are to make sense.

4. SUFFICIENT CONDITIONS FOR CONTINUITY

In this section we obtain sufficient conditions for the continuity a.s. of the sample paths of certain ξ-radial processes in terms of metric entropy. This section parallels Section 3 [MP2] where this was done for p-stable processes, $1 < p < 2$. However, because we are working with a more general class of processes we are confronted by numerous technical difficulties that were not present in the earlier work.

Let N denote the positive integers (excluding zero) and let $\delta: N \to \mathbb{R}$, $\delta(0) = 0$, $\delta(1) > 0$, be strictly increasing and regularly varying at infinity with index $1/2 < q < 1$. Let I be an index set. We will denote by $\ell_{\delta,\infty}(I)$ the space of all families $(\alpha_i)_{i \in I}$ of complex numbers such that

$$\sup_{t>0} t\delta(\text{card}\{i \in I: |\alpha_i| > t\}) < \infty$$

and we define

$$\|(\alpha_i)_{i \in I}\|_{\delta,\infty} = \sup_{t>0} t\delta(\text{card}\{i \in I: |\alpha_i| > t\}).$$

For any family $(\alpha_i)_{i \in I}$ of complex numbers tending to zero at infinity, we can define a sequence $(\alpha_n^*)_{n \in N}$ which is the non-increasing rearrangement of $(|\alpha_i|)_{i \in I}$. It is easy to check that

$$(4.1) \qquad \|(\alpha_i)_{i \in I}\|_{\delta,\infty} = \sup_{n \geq 1} \delta(n)\alpha_n^*.$$

For $1 < p < 2$, the standard $\ell_{p,\infty}(N)$ spaces are obtained when $\delta(n) = n^{1/p}$. As in these cases $\| \quad \|_{\delta,\infty}$ is equivalent to a norm on $\ell_{\delta,\infty}(I)$. (To see this we first note that $\sup_{n \geq 1} \dfrac{\delta(n)}{n} \sum_{k=1}^{n} \alpha_k^*$ is clearly a norm on those families $(\alpha_i)_{i \in I}$ for which $\sup_{n \geq 1} \dfrac{\delta(n)}{n} \sum_{k=1}^{n} \alpha_k^* < \infty$. We next

60

observe that

$$\sup_n \delta(n)\alpha_n^* \leq \sup_n \frac{\delta(n)}{n} \sum_{k=1}^{n} \alpha_k^*$$

$$\leq \sup_{n\geq 1} \left(\frac{\delta(n)}{n} \sum_{k=1}^{n} \frac{1}{\delta(k)} \right) \sup_{k\geq 1} \delta(k)\alpha_k^*$$

$$\leq \text{Const.} \sup_{k\geq 1} \delta(k)\alpha_k^* \,,$$

where we use the fact that $\delta(n)$ is regularly varying of index less than one to see that $\sup_{n\geq 1} \frac{\delta(n)}{n} \sum_{k=1}^{n} \frac{1}{\delta(k)} < \infty$.)

We impose on $\delta(n)$ an additional condition namely that the function

$$(4.2) \qquad\qquad q(n) = \frac{n}{\delta(n)}, \quad n \in N$$

is strictly increasing and concave. For $1 \leq x < \infty$ and $n \in N$ we define

$$(4.3) \qquad q(x) = \begin{cases} q(n); & x = n \\ \\ q(n+1)(x-n) + (n+1-x)q(n); & n < x < n+1 \end{cases} \quad .$$

Clearly, $q^{-1}(x)$ is convex $1 \leq x \leq \infty$, and we extend this function so that $q^{-1}(x)$ is convex and strictly increasing for $x \geq 0$, with $q^{-1}(0) \geq 0$.

Let us denote by Ψ_q the function

$$(4.4) \qquad\qquad \Psi_q(x) = \exp\left[q^{-1}(x) - q^{-1}(0) \right] - 1, \quad x \geq 0 \ .$$

We will be interested in the Orlicz space $L^{\Psi_q}(dP)$ formed by measurable functions $f: \Omega \to \mathbb{C}$ for which there is a $c > 0$ such that

$$E\Psi_q\left(\left| \frac{f}{c} \right| \right) < \infty$$

equipped with the norm

$$(4.5) \qquad\qquad \| f \|_{\Psi_q} = \inf\{ c > 0: \ E\Psi_q\left(\left| \frac{f}{c} \right| \right) \leq 1 \} \ .$$

As usual let $\{\varepsilon_i\}_{i \in I}$ be a Rademacher sequence on some probability space. It is easy to see since $\delta(n)$ is regularly varying of index $1/2 < q < 1$, that

$$\ell_{\delta,\infty}(I) \subset \ell_2(I)$$

(where $(\alpha_i)_{i \in I} \in \ell_2(I)$ iff $\sum_{i \in I} |\alpha_i|^2 < \infty$.) Therefore if $(\alpha_i)_{i \in I} \in \ell_{\delta,\infty}(I)$ the series $S = \sum_{i \in I} \varepsilon_i \alpha_i$ converges a.s.

The next lemma, which extends Lemma 3.1 [MP2] has essentially the same proof as its predecessor.

__Lemma 4.1.__ If $(\alpha_i)_{i \in I}$ belongs to $\ell_{\delta,\infty}(I)$ then S belongs to $L^{\Psi_q}(dP)$ and we have

(4.6) $k^{-1} \| (\alpha_i)_{i \in I} \|_{\delta,\infty} \leq \| S \|_{\Psi_q} \leq k \| (\alpha_i)_{i \in I} \|_{\delta,\infty}$

where k is a constant depending only on the function δ.

__Proof.__ Clearly we may assume that $I = N$ and $(|\alpha_n|)_{n \in N}$ is non-increasing. We first show the right side of (4.6). Assume $\sup_n \delta(n)|\alpha_n| \leq 1$ and let $S_n = \sum_{k=1}^{n} \varepsilon_k \alpha_k$. Then for all $u > 0$

(4.7) $P(|S| > 2u) \leq P(|S_n| > u) + P(|S - S_n| > u)$.

We choose $u = \sum_{k=1}^{n} \frac{1}{\delta(k)}$ which makes $P(|S_n| > u) = 0$. By a well known estimate for subgaussian series (cf. [K] p. 43) we have

(4.8) $P(|S - S_n| > u) \leq 2 \exp\{- \frac{u^2}{2}(\sum_{k > n} \delta(k)^{-2})^{-1}\}$.

Since $\delta(k)$ is regularly varying of index $1/2 < q < 1$ there exist constants C_1 and C_2 such that

(4.9) $\sum_{k=1}^{n} \delta(k)^{-1} \geq C_1 \frac{n}{\delta(n)} = C_1 q(n)$

and

(4.10)
$$\sum_{k=n+1}^{\infty} \delta(k)^{-2} \le C_2 \frac{n}{(\delta(n))^2}.$$

Therefore by (4.8) – (4.10) we have

(4.11)
$$P(|S-S_n| > C_1 q(n)) \le \exp -C_3 n$$

where $C_3 = C_1^2(2C_2)$. Therefore using (4.7) we have that

(4.12)
$$P(|S| > 2C_1 n) \le \exp -C_3 q^{-1}(n) .$$

To complete this portion of the proof we must show that there exists an a such that

$$E \exp q^{-1}\left(\left|\frac{S}{a}\right|\right) \le 2 \exp q^{-1}(0) .$$

By the dominated convergence theorem the existence of such a number a will follow if we can show that there exists an a_1 such that

(4.13)
$$E \exp q^{-1}\left(\left|\frac{S}{a_1}\right|\right) < \infty .$$

By standard arguments employing integration by parts (4.13) is satisfied iff

(4.14)
$$\int_0^{\infty} P(|S| > u) \, d\left(\exp q^{-1}\left(\left|\frac{u}{a_1}\right|\right)\right) < \infty .$$

The integral in (4.14)

(4.15)
$$= \int_0^{\infty} P(|S| > 2C_1 u) \, d\left(\exp q^{-1}\left(\left|\frac{2C_1 u}{a_1}\right|\right)\right)$$

$$\le \sum_{n=0}^{\infty} P(|S| > 2C_1 n) \exp q^{-1}\left(\left|\frac{2C_1(n+1)}{a_1}\right|\right)$$

$$\le \sum_{n=0}^{\infty} \exp -\left[C_3 q^{-1}(n) - q^{-1}\left(\left|\frac{2C_1(n+1)}{a_1}\right|\right)\right]$$

where at the last step we use (4.12). Since q^{-1} is regularly varying it is easy to see that there exists an a_1 for which (4.15) and hence (4.14) and (4.13) are finite. This completes the proof of the right hand inequality in (4.6).

The left side of (4.6) is simple. Note that $\sum\limits_{k=1}^{n} \varepsilon_k = n$ with probability 2^{-n}. Therefore

$$E \exp q^{-1}\left\{\frac{\left|\sum\limits_{k=1}^{n} \varepsilon_k\right|}{\delta(n)}\right\} \geq 2^{-n} E \exp q^{-1}\left\{\frac{n}{\delta(n)}\right\} = \left(\frac{e}{2}\right)^n.$$

Let n_0 be such that $\left(\frac{e}{2}\right)^n \geq 2 \exp q^{-1}(0)$, $n \geq n_0$. This implies that for $n \geq n_0$

(4.16) $$\left\|\sum_{k=1}^{n} \varepsilon_k\right\|_{\Psi_q} \geq \delta(n).$$

Therefore, we have for $n \geq n_0$

(4.17) $$\left\|\sum_{k=1}^{\infty} \alpha_k \varepsilon_k\right\|_{\Psi_q} \geq \left\|\sum_{k=1}^{n} \alpha_k \varepsilon_k\right\|_{\Psi_q}$$

$$\geq \frac{1}{2} \left\|\sum_{k=1}^{n} \varepsilon_k\right\|_{\Psi_q} \inf_{k \leq n} \alpha_k \geq \frac{1}{2} \delta(n) \alpha_n^*.$$

Note that the first inequality in (4.17) follows by convexity. The second inequality follows from the contraction principle (see e.g. [MP1], p. 45, which works for more general norms that $\left(E\|\ \|^p\right)^{1/p}$) and the last inequality follows from (4.16). By (4.17) we see that

(4.18) $$\|S\|_{\Psi_q} \geq 1/2 \sup_{n \geq n_0} \delta(n) \alpha_n^*.$$

Of course the left side of (4.6) is true when $|\alpha_1| = 0$. Thus we can assume, by homogeneity, that $|\alpha_1| = 1$. In this case $\|S\|_{\Psi_q} > 0$ so there exists a $b > 0$ such that

(4.19) $$\|S\|_{\Psi_q} \geq b \sup_{n \leq n_0} \delta(n).$$

Combining (4.18) and (4.19) we get

$$\|S\|_{\Psi_q} \geq (b \wedge 1/2)\Big[\sup_{n \leq n_0} \delta(n) \vee \sup_{n \geq n_0} \delta(n)a_n^*\Big]$$

which implies the left side of (4.6). This completes the proof of the
lemma.

   The next lemma is a well known variant of Dudley's Theorem
(cf. [NN]).

Lemma 4.2. Let (T,d) be a compact metric space and let $q(x)$,
$1 \leq x < \infty$, $q^{-1}(x)$, $0 \leq x < \infty$ and $\Psi_q(x)$, $x \geq 0$ be as defined in (4.3)
and (4.4). Extend $q(x)$ to $0 \leq x < 1$ by setting it equal to
$(q^{-1}(x))^{-1}$, $0 < x < 1$. Assume that

$$J(q,d) \equiv \int_0^{\hat{d}} q(\log N(T,d;\varepsilon))d\varepsilon < \infty$$

where $\hat{d} = \sup_{s,t \in T} d(s,t)$. Then any random process $(X(t))_{t \in T}$ in $L^{\Psi_q}(dP)$
satisfying

(4.20) $\forall s,\ t \in T$ $\|X(s) - X(t)\|_{\Psi_q} \leq d(s,t)$

has a version with continuous sample paths and

$$E \sup_{s,t \in T} |X(s) - X(t)| \leq C(J(q,d) + \hat{d})$$

where C is a constant depending only on q.

   The reader is referred to [MP1] p. 25 where a proof of this result is
given in the case $q(x) = x^{1/2}$, $0 \leq x < \infty$. The case considered here is
entirely similar since (4.20) implies

$$P\left\{\left|\frac{X(s) - X(t)}{d(s,t)}\right| > u\right\} \leq 2 \exp q^{-1}(0) \exp - q^{-1}(u)\ ,\quad \forall u > 0\ .$$

For a more general discussion see [P], [F2].

We now come to the crucial part of this section. We begin by considering the expected value of the norm in $\ell_{\delta,\infty}(N)$ for sequences $\{Z_n\}_{n\in N}$ of random variables.

<u>Lemma 4.3</u>: Let $\{Z_n\}_{n\in N}$ be non-negative independent random variables and let $\theta: N \to R^+$, $\theta(1) = 1$, be a concave increasing real valued function such that for any $\varepsilon > 0$ there exists an integer $n_0(\varepsilon)$ for which

$$(4.21) \qquad \theta\big(\big[(nL_2n)^{1/2}\big] + 1\big) \leq \varepsilon\theta(n), \quad \forall n \geq n_0(\varepsilon).$$

(Here [ ] denotes integral part and $L_2n = \max[1, \log \log n]$.) Set $\theta(0) = 0$; then there exists a constant C_θ, depending only on θ, such that

$$(4.22) \qquad \frac{1}{2}\Big\{ \sup_{t>0} Et\theta\Big(\sum_{n=1}^\infty I_{[Z_n>t]}\Big) + E \sup_{n\geq 1} Z_n\Big\}$$

$$\leq E \sup_{n\geq 1} \theta(n)Z_n^*$$

$$\leq \sup_{t>0} Et\theta\Big(\sum_{n=1}^\infty I_{[Z_n>t]}\Big) + C_\theta E \sup_{n\geq 1} Z_n$$

where $\{Z_n^*\}_{n\in N}$ denotes a non-decreasing rearrangement of $\{Z_n^*\}_{n\in N}$ and $I_{[A]}$ denotes an indicator function of the set A. (Of course the reader will recall that $E \sup_{n\geq 1} \theta(n)Z_n^*$ is, by definition, $E\|(Z_n)_{n\in N}\|_{\theta,\infty}$ as defined in (4.1)).

<u>Proof</u>: Since θ is concave on N (the set of positive integers excluding zero) and $\theta(0) = 0$ we have that for integers $a, b \geq 0$

$$(4.23) \qquad \theta(|a-b|) \geq |\theta(a) - \theta(b)|$$

and

$$(4.24) \qquad \theta(|a-b|) \leq |\theta(a) + \theta(b)|.$$

Note that

(4.25)
$$E \sup_{n \geq 1} \theta(n) Z_n^* = E \sup_{t>0} t\theta\Big(\sum_{n=1}^{\infty} I_{[Z_n>t]} \Big).$$

Let $\|\cdot\|$ indicate $\sup_{t>0} |\cdot|$ and let $\{Z_n'\}_{n \in N}$ be an independent copy of $\{Z_n\}_{n \in N}$. Then by convexity

(4.26)
$$E\Big\| t\theta\Big(\sum_{n=1}^{\infty} I_{[Z_n>t]} \Big) - t\theta\Big(\sum_{n=1}^{\infty} I_{[Z_n'>t]} \Big)\Big\|$$

$$\geq E\Big\| t\theta\Big(\sum_{n=1}^{\infty} I_{[Z_n>t]} \Big)\Big\| - \Big\| Et\theta\Big(\sum_{n=1}^{\infty} I_{[Z_n'>t]} \Big)\Big\|.$$

Furthermore by (4.23) and (4.24) the first term in (4.26)

(4.27)
$$\leq E\Big\| t\theta\Big(\Big| \sum_{n=1}^{\infty} \big(I_{[Z_n>t]} - I_{[Z_n'>t]} \big) \Big| \Big)\Big\|$$

$$= E\Big\| t\theta\Big(\Big| \sum_{n=1}^{\infty} \varepsilon_n \big(I_{[Z_n>t]} - I_{[Z_n'>t]} \big) \Big| \Big)\Big\|$$

$$\leq 2E\Big\| t\theta\Big(\Big| \sum_{n=1}^{\infty} \varepsilon_n I_{[Z_n>t]} \Big| \Big)\Big\|$$

where $\{\varepsilon_n\}_{n \in N}$ is a Rademacher sequence independent of $\{Z_n\}_{n \in N}$ and $\{Z_n'\}_{n \in N}$. Thus by (4.25), (4.26) and (4.27) we have

(4.28)
$$E \sup_{n \geq 1} \theta(n) Z_n^* \leq \Big\| Et\theta\Big(\sum_{n=1}^{\infty} I_{[Z_n>t]} \Big)\Big\|$$

$$+ 2E\Big\| t\theta\Big(\Big| \sum_{n=1}^{\infty} \varepsilon_n I_{[Z_n>t]} \Big| \Big)\Big\|.$$

We will now concern ourselves with the last term in (4.28). Let $\pi: N \to N$ be a permutation such that $Z_{\pi(k)} = Z_k^*$. Then

(4.29)
$$\Big\| t\theta\Big(\Big| \sum_{n=1}^{\infty} \varepsilon_n I_{[Z_n>t]} \Big| \Big)\Big\| = \sup_{n \geq 1} (Z_n^*)\theta\Big(\Big| \sum_{k=1}^{n} \varepsilon_{\pi(k)} \Big| \Big).$$

Now let $\{Z_n\}_{n \in N}$ be defined on the probability space (Ω, F, P) and $\{\varepsilon_n\}_{n \in N}$ be defined on the probability space (Ω', F', P'). Let $\omega \in \Omega$ and

$\omega' \in \Omega'$ and let E_ω and $E_{\omega'}$ denote the corresponding expectation operators. Then

$$(4.30) \quad E \sup_{n \geq 1} (Z_n^*)\theta \Big(\Big| \sum_{k=1}^{n} \varepsilon_{\pi(k)} \Big| \Big) = E_\omega E_{\omega'} \sup_{n \geq 1} (Z_n^*(\omega))\theta \Big(\Big| \sum_{k=1}^{n} \varepsilon_{\pi(k)}(\omega') \Big| \Big).$$

Now, even though $\pi(k)$ depends on $\omega \in \Omega$, the fact that $\{Z_n\}_{n \in N}$ and $\{\varepsilon_n\}_{n \in N}$ are independent implies that (4.30)

$$(4.31) \qquad\qquad = E_\omega E_{\omega'} \sup_{n \geq 1} Z_n^*(\omega)\theta \Big(\Big| \sum_{k=1}^{n} \varepsilon_k(\omega') \Big| \Big) .$$

It follows from the law of the iterated logarithm there exists a set $\overline{\Omega}' \subset \Omega'$, $P(\overline{\Omega}') = 1$, such that

$$(4.32) \qquad\qquad \sup_{n \geq 1} \Big| \sum_{k=1}^{n} \varepsilon_k(\omega') \Big| (nL_2 n)^{-1/2} \leq M(\omega'), \quad \forall \omega' \in \overline{\Omega}'$$

where $M(\omega')$ is a random variable with all moments finite, [K2]. Without loss of generality we will assume that $M(\omega') \geq 1$.

Let us also note that since $\theta(n)$, $n \geq 1$ is concave there exists an integer n_1 such that for all $n \geq n_1$ and constants $a \geq 1$

$$(4.33) \qquad\qquad \theta([an] + 1) \leq a\theta(n+1).$$

In (4.21) choose ε so that $\varepsilon E_{\omega'} M(\omega') \leq 1/4$. Then by (4.32) and (4.33) for $n_2 \geq n_1$ we have

$$(4.34) \quad E_{\omega'} \theta \Big(\Big| \sum_{k=1}^{n} \varepsilon_k(\omega') \Big| \Big) \leq E_{\omega'} \theta \big([M(\omega')(nL_2 n)^{1/2}] + 1 \big)$$

$$\leq E_{\omega'} M(\omega') \theta \big([(nL_2 n)^{1/2}] + 1 \big) \leq \frac{1}{4} \theta(n).$$

By (4.29), (4.30), (4.31) and (4.34) we have

$$(4.35) \quad E\| t\theta\Big(\Big|\sum_{n=1}^{\infty} \epsilon_n I_{[Z_n>t]}\Big|\Big)\| \leq E_\omega E_{\omega'} \sup_{n\leq n_0} Z_n^*(\omega)\theta\Big(\Big|\sum_{k=1}^{n} \epsilon_k(\omega')\Big|\Big)$$

$$+ E_\omega E_{\omega'} \sup_{n\geq n_0} Z_n^*(\omega)\theta\Big(\Big|\sum_{k=1}^{n} \epsilon_k(\omega')\Big|\Big)$$

$$\leq E Z_1^*\theta(n_0) + \epsilon E_\omega, M(\omega')E_\omega \sup_{n\geq n_0} Z_n^*\theta(n)$$

$$\leq \theta(n_0) E \sup_{n\geq 1} Z_n + \frac{1}{4} E \sup_{n\geq 1} Z_n^*\theta(n) .$$

Substituting (4.35) in (4.28) we get the right side of (4.22). Using (4.25) it is easy to see that the lower bound in (4.22) is also valid.

In the next lemma we show that it is enough to consider $Et\theta\Big(\sum_{n=1}^{\infty} I_{[Z_n>t]}\Big)$ for small values of t. The awkwardness encountered in the proof is because θ is defined only on the integers and in extending it to $x \in [0,\infty)$ we need to keep $\theta(x) = 0$ for $x \in [0,1)$.

Lemma 4.4: Let $\{Z_n\}_{n\in N}$ and θ be as in Lemma 4.3. Extend θ to a function from $[0,\infty) \to \mathbb{R}^+$ such that θ is a concave function on $x \geq 1$ and $\theta(x) = 0$, $0 \leq x < 1$. Let

$$(4.36) \qquad\qquad a(t) = \sum_{n=1}^{\infty} P(Z_n > t)$$

and

$$(4.37) \qquad\qquad H(x) = \frac{x}{1 - e^{-x}}$$

Then, for $0 < t_0 < \infty$

$$(4.38) \quad \sup_{t>0} Et\theta\Big(\sum_{n=1}^{\infty} I_{[Z_n>t]}\Big) \leq \sup_{t<t_0} Et\theta\Big(\sum_{n=1}^{\infty} I_{[Z_n>t]}\Big) + \theta\big(H(a(t_0))\big)E \sup_{n\geq 1} Z_n .$$

Proof: We first observe that

$$(4.39) \qquad E\theta\left(\sum_{n=1}^{\infty} I_{[Z_n>t]}\right) \le \theta\left(\frac{\sum_{n=1}^{\infty} P(Z_n>t)}{P\left(\sup_{n\ge1} Z_n > t\right)}\right) P\left(\sup_{n\ge1} Z_n>t\right) \quad .$$

To verify (4.39) we assume $P\left(\sup_{n\ge1} Z_n > t\right) > 0$ (If not we evaluate the left side of (4.39) to be zero). We have

$$E\theta\left(\sum_{n=1}^{\infty} I_{[Z_n>t]}\right) = \int_1^{\infty} \theta(u)dP\left(\sum_{n=1}^{\infty} I_{[Z_n>t]} \le u\right)$$

$$= \beta_t \int_1^{\infty} \theta(u)d\nu(u)$$

where

$$(4.40) \qquad \beta_t = P\left(\sup_{n\ge1} Z_n > t\right)$$

and

$$\nu(u) = \beta_t^{-1} P\left(\sum_{n=1}^{\infty} I_{[Z_n>t]} \le u\right) .$$

Note that ν is a probability measure on $[1,\infty)$ and since θ is concave on $[1,\infty)$ we can use Jensen's inequality to obtain

$$(4.41) \qquad E\theta\left(\sum_{n=1}^{\infty} I_{[Z_n>t]}\right) \le \beta_t \theta\left(\int_1^{\infty} ud\nu(u)\right)$$

$$\le \beta_t \theta\left(\frac{\sum_{n=1}^{\infty} P(Z_n>t)}{\beta_t}\right)$$

since

$$\int_1^{\infty} ud\nu(u) = \beta_t^{-1} \int_1^{\infty} udP\left(\sum_{n=1}^{\infty} I_{[Z_n>t]} \le u\right)$$

$$= \beta_t^{-1} E\left(\sum_{n=1}^{\infty} I_{[Z_n>t]}\right).$$

Note that (4.41) is the same as (4.39).

It is easy to see that

(4.42)
$$P\left(\sup_{n\geq 1} Z_n > t\right) \geq 1 - \exp - \sum_{n=1}^{\infty} P(Z_n > t) \ .$$

Thus, by (4.42)

(4.43)
$$\sup_{t\geq t_0} \frac{\sum_{n=1}^{\infty} P(Z_n > t)}{P\left(\sup_{n\geq 1} Z_n > t\right)} \leq \sup_{t\geq t_0} \frac{a(t)}{1-e^{-a(t)}} = H(a(t_0))$$

since $H(x)$ is increasing for $x \geq 0$. Using (4.39), (4.43) and Chebyshev's inequality we get

$$\sup_{t\geq t_0} Et\theta\left(\sum_{n=1}^{\infty} I_{[Z_n>t]}\right) \leq \sup_{t\geq t_0} \theta\left(\frac{\sum_{n=1}^{\infty} P(Z_n>t)}{P\left(\sup_{n\geq 1} Z_n>t\right)}\right) \sup_{t>0} tP\left(\sup_{n\geq 1} Z_n>t\right)$$

$$\leq \theta(H(a(t_0)))E \sup_{n\geq 1} Z_n$$

which gives us (4.38).

In the next lemma we relate the expected value of a suitably normalized sequence of i.i.d random variables to an Orlicz space norm (Actually the lemma is slightly more general than this.)

Lemma 4.5. Let $\kappa: [1,\infty) \to R^+$, $\kappa(1) > 0$ be a continuous non-decreasing regularly varying function of index $1/p$, $p > 1$. Define $\kappa^{-1}(x) = \inf\{n: \kappa(n) > x\}$ and extend κ^{-1} by setting $\kappa^{-1}(x) = 1$, $0 \leq x \leq \kappa(1)$. Let $v: \Omega \to R^+$ be a random variable satisfying $E\kappa^{-1}\left(\left|\frac{v}{a}\right|\right) < \infty$ for some $a < \infty$ and define

$$\|v\|_{\kappa^{-1}} = \inf\left\{c: E\kappa^{-1}\left(\left|\frac{v}{c}\right|\right) \leq 2\right\} \ .$$

Let $\{v_n\}_{n\in N}$ be independent identically distributed copies of v, then

(4.44)
$$E \sup_{n} \frac{v_n}{\kappa(n)} \leq C_\kappa \|v\|_{\kappa^{-1}}$$

where C_κ is a constant depending only on κ. (Note that if κ^{-1} is convex $\| \ \|_{\kappa^{-1}}$ is an Orlicz space norm. We don't need to assume κ^{-1}

convex, nevertheless $\| \ \|_{\kappa^{-1}}$ is well defined.)

<u>Proof</u>: Since $\| \ \|_{\kappa^{-1}}$ is homogeneous, without loss of generality we can assume that $\kappa(1) = 1$. Let us also assume that κ is strictly increasing on $[1,\infty)$. In this case κ^{-1} is continuous. Note that

$$P\left(\sup_{n\geq 1} \frac{v_n}{\kappa(n)} > c\right) = P\left(\sup_{n\geq 1} \kappa^{-1}\left(\frac{v_n}{c}\right) I_{\left[\frac{v_n}{c} > 1\right]} > n\right)$$

$$\leq 1 \wedge \sum_{j=1}^{\infty} P\left(\kappa^{-1}\left(\frac{v}{c}\right) I_{\left[\frac{v}{c} > 1\right]} > j\right)$$

$$\leq 1 \wedge E\kappa^{-1}\left(\frac{v}{c}\right) I_{\left[\frac{v}{c} > 1\right]}$$

$$= 1 \wedge \int_c^{\infty} \kappa^{-1}\left(\frac{u}{c}\right) dP(v \leq u).$$

Therefore

(4.45) $$E \sup_{n\geq 1} \frac{v_n}{\kappa(n)} \leq \|v\|_{\kappa^{-1}} + \int_{\|v\|_{\kappa^{-1}}}^{\infty} \int_c^{\infty} \kappa^{-1}\left(\frac{u}{c}\right) dP(v \leq u) dc$$

$$= \|v\|_{\kappa^{-1}} + \int_{\|v\|_{\kappa^{-1}}}^{\infty} \int_{\|v\|_{\kappa^{-1}}}^{u} \kappa^{-1}\left(\frac{u}{c}\right) dc\, dP(v \leq u).$$

$$= \|v\|_{\kappa^{-1}} + \int_{\|v\|_{\kappa^{-1}}}^{\infty} u \int_1^{u/\|v\|_{\kappa^{-1}}} \frac{\kappa^{-1}(\lambda)}{\lambda^2} d\lambda\, dP(v \leq u).$$

Since κ^{-1} is regularly varying of index $1 < p$

$$\lim_{x\to\infty} \int_1^x \frac{\kappa^{-1}(\lambda)}{\lambda^2} d\lambda = \infty$$

and furthermore there exists a constant D_κ (depending on κ^{-1} or equivalently κ) such that

$$\int_1^x \frac{\kappa^{-1}(\lambda)}{\lambda^2} d\lambda \leq D_\kappa \frac{\kappa^{-1}(x)}{x}$$

(see e.g. [K1] or [F] pg. 273). Therefore we have

$$u \int_1^{u/\|v\|_{\kappa^{-1}}} \frac{\kappa^{-1}(\lambda)}{\lambda^2} \, d\lambda \le D_\kappa \|v\|_{\kappa^{-1}} \, \kappa^{-1}\left(\frac{u}{\|v\|_{\kappa^{-1}}}\right)$$

and the double integral on the last line of (4.45)

(4.46)
$$\le D_\kappa \|v\|_{\kappa^{-1}} \int_{\|v\|_{\kappa^{-1}}}^\infty \kappa^{-1}\left(\frac{u}{\|v\|_{\kappa^{-1}}}\right) dP(v \le u)$$

$$\le 2D_\kappa \|v\|_{\kappa^{-1}}$$

because

$$\int_{\|v\|_{\kappa^{-1}}}^\infty \kappa^{-1}\left(\frac{u}{\|v\|_{\kappa^{-1}}}\right) dP(v \le u) \le E\kappa^{-1}\left(\frac{v}{\|v\|_{\kappa^{-1}}}\right) \le 2$$

(For the final inequality we use the fact that κ^{-1} is continuous).
Combining (4.45) and (4.46) we get (4.44) in the case when κ is strictly
increasing on $[1,\infty)$.

If κ is not strictly increasing we can find a function
$\alpha: [1,\infty) \to \mathbb{R}^+$, with $\alpha(1) = 1$ and $\lim_{x \to \infty} \alpha(x) = 1+\varepsilon$ which is strictly
increasing so that $(\alpha\kappa)(x) \equiv \alpha(x)\kappa(x)$ is strictly increasing with
$(\alpha k)(1) = 1$. Clearly

(4.47)
$$E \sup_n \frac{v_n}{\kappa(n)} \le (1+\varepsilon) \, E \sup_n \frac{v_n}{(\alpha\kappa)(n)} \le C\|v\|_{(\alpha\kappa)^{-1}}$$

by the proof above which applies to the function $(\alpha\kappa)$, where C is a
constant independent of $\{v_n\}_{n \in N}$. Since $(\alpha\kappa)^{-1} \le \kappa^{-1}$ we now have (4.44)
in the generality stated in the hypothesis of this lemma.

We now come to the main theorem of this section. Recall that
$\theta: N \to \mathbb{R}^+$, $\theta(1) = 1$ is concave and strictly increasing and satisfies
(4.21) a regularity condition. We extend θ to a funciton on $[0,\infty)$
satisfying $\theta(x) = 0$, $0 \le x < 1$ and $\theta(x)$ concave $x \in [1,\infty)$. Define

$\theta^{-1}(x) = \inf\{u: \theta(u) > x\}$. $\theta^{-1}(x)$ is a convex non-decreasing function on $[0,\infty)$ with $\theta^{-1}(0) = 1$. In Lemma 4.5 we introduced $\kappa: [1,\infty) \rightarrow \mathbb{R}^+$ a continuous non-decreasing regularly varying function of index $1/p$, $p > 1$ and we defined $\kappa^{-1}(x) = \inf\{u: \kappa(u) > x\}$ extended to a function on $[0,\infty)$ by setting $\kappa^{-1}(x) = 1$, $0 \leq x \leq \kappa(1)$. Let us also recall that in the Introduction of this paper we introduced $\Phi: \mathbb{R}^+ \rightarrow \mathbb{R}^+$, $\Phi(0) = 0$ to be a convex, non-decreasing function. All of these function play a role in the next theorem .

<u>Theorem 4.6.</u> Let κ, κ^{-1}, θ and Φ be as above with the additional conditions: There exists an $x_0 \geq 0$ and a strictly increasing real valued function T such that

$$(4.48) \qquad\qquad \kappa^{-1}(x) \leq k_1 T(x), \quad x \geq x_0$$

for some constant k_1 where the function T satisfies

$$(4.49) \qquad\qquad T(xy) \leq k_2 \theta^{-1}(x)\Phi(y), \quad \forall x \geq 1, \quad y \geq 1 .$$

Let v be a non-negative random variable satisfying $E\kappa^{-1}(v) < \infty$ and let $\{v_n\}_{n\in N}$ be independent copies of v. Let $\left(\left(\frac{v_n}{\kappa(n)}\right)^*\right)_{n\in N}$ denote a non-decreasing rearrangment of $\left(\frac{v_n}{\kappa(n)}\right)_{n\in N}$. Then we have

$$(4.50) \qquad\qquad E \sup_{n \geq 1} \theta(n)\left(\frac{v_n}{\kappa(n)}\right)^* \leq C\|v\|_\Phi$$

where C is a constant independent of $\{v_n\}_{n\in N}$ and $\|v\|_\Phi$ is as defined in (4.5).

Before proving this theorem we will obtain a Corollary which gives some nice forms of (4.50).

<u>Corollary 4.7:</u> Let $\Phi: \mathbb{R}^+ \rightarrow \mathbb{R}^+$, $\Phi(0) = 0$, $\Phi(1) = 1$, be a convex strictly increasing function that is regularly varying at infinity with index $p > 1$, let v be a non-negative random variable satisfying $E\Phi(v) < \infty$ and let $\{v_n\}_{n\in N}$ be independent copies of v.

Suppose that there exists a constant k such that

$$(4.51) \qquad \Phi(xy) \leq k\Phi(x)\Phi(y), \quad \forall x \geq 1, \quad y \geq 1,$$

then

$$(4.52) \qquad E \sup_{n \geq 1} \Phi^{-1}(n)\Big(\frac{v_n}{\Phi^{-1}(n)}\Big)^* \leq C\|v\|_\Phi .$$

Suppose that there exists a constant k such that

$$(4.53) \qquad \Phi(xy) \leq kx^P\Phi(y), \quad \forall x \geq 1, \quad y \geq 1 ,$$

then both

$$(4.54) \qquad E \sup_{n \geq 1} n^{1/p}\Big(\frac{v_n}{\Phi^{-1}(n)}\Big)^* \leq C\|v\|_\Phi$$

and

$$(4.54a) \qquad E \sup_{n \geq 1} \Phi^{-1}(n)\Big(\frac{v_n}{\Phi^{-1}(n)}\Big)^* < C\|v\|_p$$

where C is a constant independent of $\{v_n\}_{n \in N}$, $\| \ \|_\Phi$ is defined as in (4.5) and we denote $\| \ \|_\Phi$ by $\| \ \|_p$ when $\Phi(x) = x^P$.

Proof of Corollary 4.7: This follows immediately from Theorem 4.6. To show that (4.51) implies (4.52) we take $\kappa^{-1}(x) = T(x) = \Phi(x)$ and $\theta(x) = \Phi^{-1}(x)$, $x \geq 1$. To show that (4.53) implies (4.54) we take $\kappa^{-1}(x) = \Phi(x) = T(x)$ and $\theta(x) = x^{1/p}$, $x \geq 1$. Lastly to show that (4.53) implies (4.54a) we set $\Phi(x)$ in Theorem 4.6 equal to x^P. Then take $T(x) = \kappa^{-1}(x) = \Phi(x)$ and $\theta^{-1}(x) = \Phi(x)$, $x \geq 1$ for the more general Φ defined in this Corollary.

Remark 4.8: For later use in Corollary 1.3 we remark that if we take $\Phi(x)$ such that

$$(4.55) \qquad \Phi(x) = x^P\big(\log(e+x)\big)^\beta, \quad x \geq x_0$$

then if $\beta \geq 0$ we can extend Φ to a convex function on $[0,\infty)$ with $\Phi(0) = 0$ such that (4.51) is satisfied. If $\beta \leq 0$ we can extend Φ to a convex function on $[0,\infty)$ such that (4.53) is satisfied.

<u>Proof of Theorem 4.6</u>: As in Lemma 4.5 we will first assume that $\kappa(x)$ is strictly increasing on $[1,\infty)$. This means that $\kappa^{-1}(x)$ is continuous on $[0,\infty)$ and that for $x \in [1,\infty)$, $\kappa^{-1}(\kappa(x)) = x$. It is easy to show that (4.48) and (4.49) imply that

$$(4.55a) \qquad\qquad \|v\|_{\kappa^{-1}} \leq C'\|v\|_{\Phi}$$

where C' is a constant depending only on κ^{-1} and Φ. Therefore by Lemma 4.5 we have

$$(4.56) \qquad\qquad E \sup_n \frac{v_n}{\kappa(n)} \leq D\|v\|_{\Phi}$$

where D is a constant depending only on κ and Φ. Thus by (4.22) in order to obtain (4.50) we need only show that

$$(4.57) \qquad\qquad \sup_{t>0} Et\theta\Big(\sum_{n=1}^{\infty} I_{[\frac{v_n}{\kappa(n)} > t]} \Big) \leq C'\|v\|_{\Phi}$$

for some constant C' independent of $\{v_n\}_{n \in N}$.

By Lemma 4.4 we have

$$(4.58) \qquad \sup_{t>0} Et\theta\Big(\sum_{n=1}^{\infty} I_{[\frac{v_n}{\kappa(n)} > t]} \Big) \leq \sup_{t \leq \|v\|_{\kappa^{-1}}} Et\theta\Big(\sum_{n=1}^{\infty} I_{[\frac{v_n}{\kappa(n)} > t]} \Big)$$

$$+ \theta\Big(H(a(\|v\|_{\kappa^{-1}}))\Big)E \sup_{n \geq 1} \frac{v_n}{\kappa(n)}$$

where H is given in (4.37) and by (4.36)

$$(4.59) \qquad a(\|v\|_{\kappa^{-1}}) = \sum_{n=1}^{\infty} P\Big(\frac{v_n}{\kappa(n)} > \|v\|_{\kappa^{-1}}\Big)$$

$$\leq \sum_{n=1}^{\infty} P\Big(\kappa^{-1}\big(\frac{v}{\|v\|_{\kappa^{-1}}}\big) > n\Big)$$

$$\leq E \kappa^{-1}\big(\frac{v}{\|v\|_{\kappa^{-1}}}\big) \leq 2 \ .$$

Therefore $H(a(\|v\|_{\kappa^{-1}})) \leq 3$. It follows from this and (4.58) and (4.56) that we can get (4.50) (for $\kappa(x)$ strictly increasing) if we can show that

$$(4.60) \qquad \sup_{t < \|v\|_{\kappa^{-1}}} Et\theta \left(\sum_{n=1}^{\infty} I_{\left[\frac{v_n}{\kappa(n)} > t\right]} \right) \leq C' \|v\|_{\Phi}$$

for a constant C' independent of $\{v_n\}_{n \in N}$.

We observe that

$$(4.61) \qquad E\kappa^{-1}\left(\frac{v}{t}\right) \geq \sum_{n=1}^{\infty} P\left(\frac{v_n}{\kappa(n)} > t\right) \geq E\kappa^{-1}\left(\frac{v}{t}\right) - 1.$$

Therefore by (4.42) and (4.60) we see that for, $0 < t \leq \|v\|_{\kappa^{-1}}$

$$(4.62) \qquad P\left(\sup_{n \geq 1} \frac{v_n}{\kappa(n)} > t\right) \geq 1 - \exp -\left(E\kappa^{-1}\left(\frac{v}{\|v\|_{\kappa^{-1}}}\right) - 1\right) \geq 1 - 1/e$$

since $E\kappa^{-1}\left(\frac{v}{\|v\|_{\kappa^{-1}}}\right) = 2$. Using (4.62) in (4.39) we have

$$(4.63) \qquad \sup_{t \leq \|v\|_{\kappa^{-1}}} tE\theta \left(\sum_{n=1}^{\infty} I_{\left[\frac{v_n}{\kappa(n)} > t\right]} \right) \leq \sup_{t \leq \|v\|_{\kappa^{-1}}} t\theta \left(\frac{\sum_{n=1}^{\infty} P\left(\frac{v_n}{\kappa(n)} > t\right)}{1 - 1/e} \right) .$$

We now use (4.61), (4.48) and (4.49) to see that

$$(4.64) \qquad \sup_{t \leq \|v\|_{\kappa^{-1}}} t\theta \left(\frac{\sum_{n=1}^{\infty} P\left(\frac{v_n}{\kappa(n)} > t\right)}{1 - 1/e} \right) \leq \sup_{t \leq \|v\|_{\kappa^{-1}}} t\theta \left(\frac{E\kappa^{-1}\left(\frac{v}{t}\right)}{1 - 1/e} \right)$$

$$\leq \sup_{t \leq \|v\|_{\kappa^{-1}}} t\theta \left(\frac{\kappa^{-1}(x_0) + k_1 ET\left(\frac{v}{t}\right)}{1 - 1/e} \right) .$$

Let us assume that C' in (4.55a) is greater than or equal to one. By (4.49) and (4.55a) we have for $t \leq \|v\|_\kappa^{-1}$ that

$$T\left(\frac{v}{t}\right) \leq k_2 \theta^{-1}\left(\frac{C'\|v\|_\Phi}{t}\right)\left[\Phi\left(\frac{v}{C'\|v\|_\Phi}\right) + \Phi(1)\right]$$

so

$$ET\left(\frac{v}{t}\right) \leq k_2'\theta^{-1}\left(\frac{C'\|v\|_\Phi}{t}\right)$$

where $k_2' = k_2(1 + \Phi(1))$, since

$$E\Phi\left(\frac{v}{C'\|v\|_\Phi}\right) \leq E\Phi\left(\frac{v}{\|v\|_\Phi}\right) = 1.$$

Thus the first term in (4.64)

$$(4.65) \qquad \leq \sup_{t \leq \|v\|_\kappa^{-1}} t\theta\left(\frac{\kappa^{-1}(x_0) + k_1 k_2'\theta^{-1}\left(\frac{C'\|v\|_\Phi}{t}\right)}{1 - 1/e}\right).$$

Recall that θ is concave on $[1,\infty)$ thus $\theta'(x) \equiv \theta(x+1)$ is concave on $[0,\infty)$ and satisfies $\theta'(a+b) \leq \theta'(a) + \theta'(b)$ and $\theta'(cd) \leq c\theta'(d)$, $c \geq 1$, $d \geq 0$. Using these inequalities we see that for $b, y \geq 1$, $a \geq 0$

$$\theta(a + by) \leq \theta(a+b) + b\theta(y) .$$

Using this we see that (4.65), with $y = \theta^{-1}\left(\frac{C'\|v\|_\Phi}{t}\right)$

$$(4.66) \qquad \leq \|v\|_\kappa^{-1} \theta\left(\frac{k^{-1}(x_0) + k_1 k_2'}{1 - 1/e}\right) + \frac{k_1 k_2' C'}{1 - 1/e} \|v\|_\Phi$$

where we use the fact that $\dfrac{C'\|v\|_\Phi}{t} \geq 1$ for $t \leq \|v\|_\kappa^{-1}$ and $\theta(\theta^{-1}(x)) = x$ for $x \geq 1$. Also without loss of generality, we assume that $k_1 k_2' \geq 1 - 1/e$. Using (4.63) - (4.66) and (4.55a) we get (4.60). This completes the proof of the theorem in the case when $\kappa(x)$ is strictly increasing on $[1,\infty)$.

As we did at the end of Lemma 4.5 we consider $(\alpha\kappa)(x) \equiv \alpha(x)\kappa(x)$ which is strictly increasing with $(\alpha\kappa)(1) = 1$ and $(\alpha\kappa)(x) < (1+\varepsilon)\kappa(x)$. Thus we have

$$E \sup_{n\geq 1} \theta(n)\left(\frac{v_n}{\kappa(n)}\right)^* \leq (1+\varepsilon) \; E \sup_{n\geq 1} \theta(n)\left(\frac{v_n}{(\alpha\kappa)(n)}\right)^* .$$

These last two inequalities complete the proof of Theorem 4.6.

We can now prove Theorem 1.1 part II.).

Proof of Theorem 1.1, part II.): Let us start with the real case. Since $\{X(t)\}_{t\in K}$ has characteristic functional (1.3) by (2.21) of Lemma 2.3 with $h_1 = \varepsilon_1$ where $\{\varepsilon_n\}_{n=1}^{\infty}$ is a Rademacher sequence we can write

$$(4.67) \qquad\qquad X(t) = \sum_{n=1}^{\infty} \varepsilon_n F^{-1}(\Gamma_n)\gamma_n(t), \quad t \in K$$

Let us recall that $\gamma \in \Gamma$ is a random variable with probability distribution given by m,

$$(4.68) \qquad\qquad F^{-1}(t) = \sup\{u: \tau[u,\infty) > t\}$$

where τ and Ψ are related by (1.2) and $\{\Gamma_j\}_{j=1}^{\infty}$ is defined in (2.9)

By the strong law of large numbers $\lim_{n\to\infty} \frac{\Gamma_n}{n} = 1$. Using this and Lemma 2.8 we see that the series in (4.67) converges uniformly a.s. iff the series

$$(4.69) \qquad\qquad Y(t) = \sum_{n=1}^{\infty} \varepsilon_n F^{-1}(n)\gamma_n(t), \quad t \in K$$

converges uniformly a.s. Indeed what we will show is that under the hypothesis of this theorem the series (4.69) converges uniformly a.s.

We now define κ and κ^{-1} as follows

$$(4.69a) \qquad\qquad \kappa(t) = \sup\{u: \tau\left[\frac{1}{u}, \infty\right) > t\}$$

and

$$\kappa^{-1}(x) = \inf\{u: \kappa(u) > x\} \ .$$

One can check that

(4.70) $$\kappa^{-1}(x) = \tau\left[\frac{1}{x}, \infty\right)$$

and that for $n \, \varepsilon \, N$

(4.71) $$\kappa(n) = \frac{1}{F^{-1}(n)} \ .$$

It follows by a Tauberian theorem (see e.g. Theorem 4, [BT]) that if $\Psi(x)$ is regularly varying at infinity with index $0 < p < 2$ then $\tau[u,\infty)$ is regularly varying at zero and

(4.71a) $$\lim_{x \to \infty} \frac{\tau\left[\frac{1}{x}, \infty\right]}{\Psi(x)} = C_p$$

where C_p is a constant depending only on the index p of regular variation of Ψ at infinity. Therefore, there exist real numbers C and x_0 such that

(4.72) $$\tau\left[\frac{1}{x}, \infty\right) \leq C\Psi(x), \qquad x \geq x_0 \ ,$$

and so, from (4.70), (4.72) and (1.14) we have that there exists real numbers x_0'' and k_1 such that

(4.73) $$\kappa^{-1}(x) \leq k_1' \Phi(x), \qquad x \geq x_0'' \ .$$

Since $\{\varepsilon_n\}_{n \varepsilon N}$ and $\{\gamma_n\}_{n \varepsilon N}$ are independent we can consider that $\{\varepsilon_n\}_{n \varepsilon N}$ is defined on the probability space (Ω', F', P') and that $\{\gamma_n\}_{n \varepsilon N}$ is defined on the probability space (Ω, F, P) and, of course, that the series in (4.69) is defined on the product probability space $(\Omega \times \Omega', F \times F', P \times P')$. Let $\omega \, \varepsilon \, \Omega$ and $\omega' \, \varepsilon \, \Omega'$. We now write the series in (4.69) as

(4.74) $$Y(t;\omega,\omega') = \sum_{n=1}^{\infty} \varepsilon_n(\omega') \frac{\gamma_n(t;\omega)}{\kappa(n)}, \qquad t \, \varepsilon \, K$$

where we also use (4.71). Now we fix $\omega \in \Omega$ and study the subgaussian series

$$(4.75) \qquad Y(t;\omega,\cdot) = \sum_{n=1}^{\infty} \varepsilon_n \frac{\gamma_n(t;\omega)}{\kappa(n)}, \quad t \in K.$$

We will show that the series in (4.75) converges uniformly a.s. for $\omega \in \overline{\Omega} \subset \Omega$ with $P(\overline{\Omega}) = 1$. This implies, by Fubini's Theorem that the series in (4.69) and consequently the one in (4.67) converges uniformly a.s., thus completing the proof of this theorem in the real case.

We introduce the random metric d_ω defined for each $\omega \in \Omega$ by

$$d_\omega(s,t) = \left\| \left(\frac{\gamma_n(s;\omega) - \gamma_n(t,\omega)}{\kappa(n)} \right)_{n \in N} \right\|_{\eta^{-1},\infty}$$

($\| \ \|_{\eta^{-1},\infty}$ is defined in the beginning of this section). Recall that an alternate expression is

$$(4.76) \qquad d_\omega(s,t) = \sup_{n \geq 1} \eta^{-1}(n) \left(\frac{\gamma_n(s;\omega) - \gamma_n(t;\omega)}{\kappa(n)} \right)^*$$

where $(\ )^*$ denotes non-increasing rearrangement. We set

$$q(n) = \frac{n}{\eta^{-1}(n)}, \quad n \in N$$

and define Ψ_q as in (4.2) and (4.3). This is possible since $q(n)$ is concave by hypothesis.

Let us note that $d_\omega(s,t)$ is always finite. To see this observe that by (4.70), (4.72) and (1.15) with $y = 1$ we have that

$$\kappa^{-1}(x) \leq C'\eta(x), \quad x \geq 1$$

and therefore

$$x \leq \kappa^{-1}(\kappa(x)) \leq C'\eta(\kappa(x)), \quad x \geq 1 \ .$$

Thus

$$\eta^{-1}(x) \leq \eta^{-1}(C'\eta(\kappa(x))) \leq C''\kappa(x), \quad x \geq 1$$

where C' and C'' are constants. (The last inequality makes use of the fact that n^{-1} is regularly varying at infinity). Therefore

$$d_\omega(s,t) \leq 2 \sup_{n \geq 1} \frac{n^{-1}(n)}{\kappa(n)} \leq 2C''.$$

By Lemma (4.1) we see that

$$(4.77) \qquad \| Y(s;\omega,\cdot) - Y(t;\omega,\cdot) \|_{\Psi_{\underset{C}{}}} \leq C d_\omega(s,t)$$

for some constant C. Therefore by Lemma 4.2 we see that $Y(t;\omega,\cdot)$ has a version with continuous paths a.s. (with respect to Ω') if

$$(4.78) \qquad \int_0^{\hat{d}_\omega} q\bigl(\log N(K,d_\omega;\epsilon)\bigr)d\epsilon < \infty$$

where $\hat{d}_\omega = \sup_{s,t \in K} d_\omega(s,t)$. Note that since n^{-1} is regularly varying at infinity $\lim_{x \to \infty} \frac{q(x)}{H(x)} = $ Const., where H is given in (1.9). Thus we can replace q by H in (4.78).

A very important point in this theorem is that $d_\omega(s,t)$ is translation invariant on G. Let us define

$$(4.79) \qquad \delta_\omega(s-t) = d_\omega(s,t)$$

Since $s, t \in K$, $\delta_\omega(u)$ is a function on $K \oplus K = \{x+y \mid x \in K,\ y \in K\}$. We will show that if

$$(4.80) \qquad \int_0^{\hat{d}} H\bigl(\log N(K,E\delta_\omega;\epsilon)\bigr)d\epsilon < \infty$$

then (4.78) holds, where $\hat{d} = \sup_{s,t \in K} E d_\omega(s,t) = \sup_{u \in K \oplus K} E\delta_\omega(u)$. We will also show that

$$(4.81) \qquad E d_\omega(s,t) \leq C \| Y(s) - Y(t) \|_\Phi = C d_{X,\Phi}(s,t)$$

for some constant C. These two assertions give us Theorem 1.1, part II.) in the real case.

That the inequality in (4.81) holds is almost exactly Theorem 4.6.

Compare (4.50), (4.76) and the first inequality in (4.81) with $\theta(n)$

replaced by $\eta^{-1}(n)$. Conditions (4.48) and (4.49) of Theorem 4.6 are the

same as (1.14), (1.15), (4.70), and (4.73). Also the conditions on

$\eta^{-1}(n)$ and $\theta(n)$ are the same and the function Φ is the same in both

cases. The only difference is that κ in Theorem 4.6 is continuous

whereas κ as defined in (4.69a) need not be. However, in showing (4.80)

we can replace κ as defined in (4.69a) by a continuous function say

$\tilde{\kappa}(t)$ which agrees with $\kappa(t)$ when t is integer valued and we can do

this in such a way that $\tilde{\kappa}^{-1} \leq \kappa^{-1}$ (i.e. $\tilde{\kappa}^{1}$ would still satisfy

(4.48)). Thus we could, if necessary, use $\tilde{\kappa}$ instead of κ. We would

then obtain (4.81) with $\tilde{\kappa}$ replacing κ (see (4.76)) but since

$\tilde{\kappa}(n) = \kappa(n)$, $n \in N$ we would also have (4.81) for κ. The second

inequality in (4.81) is just a statement of different notation that we

have used for Orlicz space norms, i.e. see (1.6) and (4.5).

We now show that (4.80) implies (4.78). This is a crucial step in

the proof but the same idea was already used in [M], [MP1], [MP2] and

explained in great detail in [MP1] in the case $H(x) = q(x) = x^{1/2}$. Let

us define

$$m_{\delta_\omega}(\varepsilon) = \mu(x \in K \oplus K \mid \delta_\omega(x) < \varepsilon)$$

where μ is Haar measure on G and δ_ω is defined in (4.79) and

$$\overline{\delta}_\omega(u) = \sup\{y: m_{\delta_\omega}(y) < u\}.$$

$\overline{\delta}_\omega$ is called the non-decreasing rearrangement of δ_ω. It follows from

Lemmas 1.1 and 1.3, Chapter II [MP1] that (4.78) holds iff

(4.82)
$$\int_0^{\overline{\delta}_\omega} q\Big(\log \frac{C}{m_{\delta_\omega}(u)}\Big)du = \int_0^{\mu_2} q\Big(\log \frac{C}{s}\Big) d\overline{\delta}_\omega(s) < \infty$$

where $\mu_2 = \mu(K \oplus K)$ and C is a constant large enough so that the log terms are greater than one. The equality in (4.82) results from the

change of variables $u = \overline{\delta_\omega(s)}$ and the fact that $\hat{\delta}_\omega = \sup_{u \in K \oplus K} \delta_\omega(u) = \hat{d}_\omega$,

(see (3.31) Chapter II [MP1] for more details). Since q and $\overline{\delta}$ are

monotone the last integral in (4.82) is finite iff

$$(4.83) \qquad\qquad \int_0^{\mu_2} \overline{\delta_\omega(s)} \; dq\left(\log \frac{C}{s}\right) < \infty \; .$$

Recall that q is concave. Let q' denote the derivative from the right

of q. The integral in (4.83)

$$(4.84) \qquad\qquad = C \int_0^{\mu_2} \frac{\overline{\delta_\omega(s)}}{s} \; q'\left(\log \frac{C}{s}\right) ds.$$

By Lemma 2.3, Chapter II [MP1]

$$(4.85) \qquad E \int_0^{\mu_2} \frac{\overline{\delta_\omega(s)}}{s} \; q'\left(\log \frac{C}{s}\right) ds \leq \int_0^{\mu_2} \frac{\overline{E\delta_\omega(s)}}{s} \; q'\left(\log \frac{C}{s}\right) ds \; .$$

Now just as we showed that the integral on the left in (4.82) is finite

iff the integral in (4.84) is finite we can show that the last integral in

(4.85) is finite iff

$$(4.86) \qquad\qquad \int_0^{\hat{\delta}_\omega} q\left(\log \frac{C}{m_{E\delta_\omega}(u)}\right) du < \infty \; .$$

Again using Lemmas 1.1 and 1.3 Chapter II [MP1] we see that (4.86) holds

iff (4.80) holds. Thus (4.80) implies (4.86) and consequently that the

first integral in (4.85) is finite. This implies that the first integral

in (4.82) is finite which implies that (4.78) holds. This completes the

proof of this theorem in the real case.

Actually the above also gives the proof in the complex case. By

Corollary 2.4 and (2.23) with $k_1 \equiv 1$ we know that the most general

complex valued ξ-radial process is of the form

$$(4.87) \qquad\qquad \sum_{n=1}^{\infty} F^{-1}(\Gamma_n) e^{i\theta_n} \gamma_n(t), \qquad t \in K \; ,$$

where $\{\theta_n\}_{n=1}^\infty$ are i.i.d. uniform random variables on $[0,2\pi]$. Recall that $F^{-1}(t) = \sup\{u:\ \tau[u,\infty) > t\}$,

$$\Psi(|\lambda|) = \int_0^\infty (\cos \lambda u-1)d\tau[u,\infty),$$

and all that is required of τ is that (2.3) holds. We showed in Lemma 2.3 that the characteristic functional of (4.87) is given by (2.20) with $\Psi_h = \Psi_\theta$ where

(4.88) $$\Psi_\theta(|\lambda|) = \frac{2}{\pi} \int_0^{\pi/2} \Psi(|\lambda \cos \theta|)d\theta,$$

(see also Lemma 2.4). Thus the function Ψ that appears in the hypothesis of Theorem 1.1 part II) must be of the form Ψ_θ as given in (4.88), i.e. Ψ_θ, by hypothesis, is regularly varying and satisfies (1.14).

Now let us consider

(4.89) $$\sum_{n=1}^\infty F^{-1}(\Gamma_n) \cos \theta_n\ \mathrm{Re}\ \gamma(t),\quad t \in K.$$

Let τ_θ be the Levy measure corresponding to Ψ_θ and define $F_\theta^{-1}(u) = \sup\{u:\ \tau_\theta[u,\infty) > t\}$. The series in (4.89) is equal in distribution to

(4.90) $$\sum_{n=1}^\infty \varepsilon_n F_\theta^{-1}(\Gamma_n)\ \mathrm{Re}\ \gamma(t),\quad t \in K$$

as we can see from (2.19) and Lemmas 2.3 and 2.4. The function Ψ_h in (2.19) that is associated with the series in (4.90) is just Ψ_θ and since Ψ_θ is regularly varying and satisfies (1.14) we conclude from the proof of the real case that (1.16) implies that the process in (4.90) has a version with continuous paths. (We can use the same proof as the proof that (4.75) has a version with continuous paths, since obviously $|\mathrm{Re}\ \gamma(s) - \mathrm{Re}\ \gamma(t)| \le |\gamma(s) - \gamma(t)|$.) We note that Ψ_θ in (4.88) remains the same if we replace $\cos \theta$ by $\sin \theta$ and the above argument doesn't change if we replace $\text{Re } \gamma(t)$ by $\text{Im } \gamma(t)$. Thus the above argument works to show continuity of each of the four real processes contained in (4.87). This completes the proof of Theorem 1.1 part II.).

Remark 4.9. In Theorem 1.1 part II.) we require that the function $\Psi(|\lambda|)$ is regularly varying at infinity with index $0 < p < 2$. In the real case it is easy to achieve this. If the Levy measure τ is regularly varying at the origin with index $-2 < p < 0$ then Ψ is regularly varying at infinity with index $0 < p < 2$ as we already remarked after (4.71). We now point out that there are many regularly varying functions Ψ that are admissible in the complex case. It follows from (2.22) of Lemma 2.3 with $k_1 = 1$ that for any Ψ as determined by (2.1) and (2.2) and g a normal random variable with mean zero and variance 1

$$\Psi_g(|\lambda|) = E_g \Psi(|g\lambda|)$$

is admissible in defining a complex valued ξ-radial process as in (2.20). We now show that if $\Psi(|\lambda|)$ is regularly varying at infinity with index $0 < p < 2$ then

(4.91)
$$\lim_{\lambda \to \infty} \frac{\Psi_g(|\lambda|)}{\Psi(|\lambda|)} = C_{g,p}$$

where $C_{g,p} > 0$ is a constant depending on p and the variance of g. In other words we can readily find regularly varying Levy transforms in the complex case.

Let $\tau[u,\infty)$ be the Levy measure corresponding to $\Psi(|\lambda|)$ then, by Lemma 2.6, we see that ξ_g, which by definition is the symmetric infinitely divisible random variable with Levy transform Ψ_g, is equal in distribution to ηg where η^2 is a positive infinitely divisible random variable with Levy measure $\tau[u^{1/2}, \infty)$. One can write the Laplace transform of η^2 in terms of the Levy measure τ as follows:

(4.92) $Ee^{-s\eta^2} = \exp - \int_0^\infty (e^{-us} - 1)d\tau[u^{1/2}, \infty) \equiv e^{-\phi(s)}$.

Using the regular variation at zero of $\tau[u^{1/2}, \infty)$ and Theorem 4 [BT] we
have that

(4.93) $$\lim_{\lambda \to \infty} \frac{\phi(\lambda^2)}{\tau[1/\lambda, \infty]} = C_p'$$

where $C_p' > 0$ is a constant depending on p. We also have, by definition
and (4.92) that

(4.94) $$e^{-\Psi_g(|\lambda|)} = Ee^{i\lambda\xi_g} = E_\eta E_g e^{i\lambda\eta g}$$

$$= E_\eta e^{-\frac{\lambda^2\sigma^2}{2}\eta^2} = e^{-\phi\left(\frac{\lambda^2\sigma^2}{2}\right)}$$

where $Eg^2 = \sigma^2$. Finally, we already noted after (4.71) that

(4.95) $$\lim_{\lambda \to \infty} \frac{\Psi(\lambda)}{\tau[1/\lambda, \infty)} = C_p''$$

where $C_p'' > 0$ is a constant depending on p. Combining (4.93), (4.94)
and (4.95) and using the fact that τ is regularly varying we get (4.91).

We will use the next result in the proof of Theorem 1.2 part II.).

<u>Theorem 4.10</u>: Let ψ, ϕ and ν satisfy the conditions of Theorem 1.2
part II.) in particular (1.27) - (1.30). Define

(4.96) $$\tau(x) = \begin{cases} 0 & 0 \le x \le 1 \\ \dfrac{1}{\nu^{-1}(\frac{1}{x})} & x \ge 1 \end{cases}.$$

Let $\{b_n\}_{n=1}^\infty$ be complex numbers and let $\{\xi_n\}_{n=1}^\infty$ be i.i.d copies of ξ
as given in (1.17). Analogous to (1.21) we define

(4.97) $$\|\{b_n\}_{n=1}^\infty\|_\phi = \inf\{c > 0: \sum_{n=1}^\infty \phi(\frac{|b_n|}{c}) \le 1\} .$$

Then we have

(4.98) $$E \sup_{n \geq 1} \tau(n)(b_n \xi_n)^* \leq C \| \{b_n\}_{n=1}^{\infty} \|_\phi \equiv C \| \{b_n\} \|_\phi$$

where   C   is a constant independent of $\{b_n\}_{n=1}^{\infty}$.

<u>Proof</u>: We note that τ and $\{|b_n \xi_n|_{n=1}^{\infty}\}$ satisfy the conditions assigned
to θ and $\{Z_n\}_{n \in N}$ in Lemmas 4.3 and 4.4. Therefore in the notation of
Lemma 4.4 we have

(4.99) $$\sup_{t>0} E\tau \left(\sum_{n=1}^{\infty} I_{[|b_n \xi_n|>t]} \right) \leq \sup_{t < \| \{b_n\} \|_\phi} E\tau \left(\sum_{n=1}^{\infty} I_{[|b_n \xi_n|>t]} \right)$$

$$+ \tau(H(a(\| \{b_n\} \|_\phi)))E \sup_{n \geq 1} |b_n \xi_n|.$$

In what follows we will consider that $x_0 = 1$ in (1.29). This can be
done without loss of generality since if $x_0 < 1$ in (1.29) we can recast
the problem with ξ replaced by $x_0 \xi$ and then (1.29) will be valid
with $x_0 = 1$. Note that

(4.100) $$P(|\xi| > \lambda) \leq C \psi\left(\frac{1}{\lambda}\right) , \forall \lambda \geq 1$$

and

(4.101) $$P(|\xi| > \lambda) \leq C_1 \phi\left(\frac{1}{\lambda}\right) \quad \forall \lambda \geq 0 .$$

The inequality in (4.100) follows from the well known estimate for the
tail of the distribution function of a random variable in terms of its
characteristic function along with the fact that $\psi(|x|)$ is regularly
varying as $|x| \to 0$. To get (4.101) we use (4.100) and (1.29) to see that

$$P(\xi > \lambda) \leq C k_1 \phi\left(\frac{1}{\lambda}\right), \quad \lambda \geq 1$$

along with the obvious observation that since ϕ is monotone

$$P(\xi > \lambda) \leq \frac{C k_1 \phi\left(\frac{1}{\lambda}\right)}{\phi(1)} , \quad \lambda \leq 1 .$$

By (4.97) and (4.101) we have that

(4.102)
$$a\left(\|\{b_n\}\|_\phi\right) = \sum_{n=1}^\infty P\left(b_n\xi_n > \|\{b_n\}\|_\phi\right)$$

$$\leq C_1 \sum_{n=1}^\infty \psi\left(\frac{|b_n|}{\|\{b_n\}\|_\phi}\right) = C_1$$

where C_1 is a constant independent of the sequence $\{b_n\}_{n=1}^\infty$.

Next let us estimate the final quantity in (4.99). We have

(4.103)
$$E \sup_{n\geq 1} |b_n\xi_n| \leq \int_0^\infty P\left(\sup_n |b_n\xi_n| > t\right)dt$$

$$\leq a + \int_a^\infty \sum_{n=1}^\infty P\left(|b_n\xi_n| > t\right)dt$$

where we take $a = \|\{b_n\}\|_\phi$. Since $|b_n| \leq |a|$ we can use (4.100) to see that the integral in (4.103)

(4.104)
$$\leq C \sum_{n=1}^\infty \int_a^\infty \psi\left(\frac{b_k}{t}\right)dt$$

$$= C \sum_{n=1}^\infty b_k \int_{a/b_k}^\infty \psi\left(\frac{1}{s}\right)ds \ .$$

Since $\psi\left(\frac{1}{s}\right)$ is regularly varying as $s \to \infty$ we can find a constant C_2 such that

$$\int_\lambda^\infty \psi\left(\frac{1}{s}\right)ds \leq C_2\lambda\psi\left(\frac{1}{\lambda}\right), \quad \forall \lambda \geq 1 \ .$$

Therefore since $a/b_k \geq 1$ we see from (1.29) that the last term in (4.104)

(4.105)
$$\leq C_2 a \sum_{n=1}^\infty \psi\left(\frac{|b_k|}{a}\right)$$

$$\leq C_2 k_1 a \sum_{n=1}^\infty \phi\left(\frac{|b_k|}{a}\right) = C_2 k_1 a \ .$$

Combining (4.103), (4.104) and (4.105) we see that

(4.106)
$$E \sup_{n\geq 1} |b_n\xi_n| \leq C\|\{b_n\}\|_\phi \ .$$

Combining (4.102) and (4.106) and using the fact that $H(x)$ is increasing

in $x > 0$ we see that the last term in (4.99) is less than or equal to

$\frac{c}{2} \| \{b_n\} \|_\phi$ where c is a constant independent of $\{b_n\}_{n=1}^\infty$. Thus in order

to prove (4.99) we need only obtain the same upper bound for the first

term to the right of the inequality in (4.99).

We begin this process by noting that if $0 < t \leq \| \{b_n\} \|_\phi$ then

$$(4.107) \qquad \sum_{n=1}^\infty P\left(|b_n \xi_n| > t \right) \leq C' \left[\nu\left(\frac{t}{\| \{b_n\} \|_\phi} \right) \right]^{-1}$$

where C' is a constant independent of $\{b_n\}_{n=1}^\infty$. This is the main step

in the proof. To obtain it we begin by using (4.101) and (1.30) to get

$$(4.108) \qquad \sum_{n=1}^\infty P\left(|b_n \xi_n| > t \right) \leq C_1 \sum_{n=1}^\infty \phi\left(\frac{|b_n|}{t} \right)$$

$$\leq \frac{C_1}{k_2} \left[\nu\left(\frac{t}{\| \{b\} \|_\phi} \right) \right]^{-1} \sum_{n=1}^\infty \phi\left(\frac{|b_n|}{\| \{b_n\} \|_\phi} \right)$$

which gives (4.107) since the final sum in (4.108) is equal to 1. Also

note that the inequality in (4.39) holds with θ and $\{Z_n\}_{n \in N}$ replaced

by τ and $\{ |b_n \xi_n| \}_{n=1}^\infty$. Therefore by (4.39) and (4.107) we have

$$(4.109) \qquad \sup_{t \leq \| \{b_n\} \|_\phi} E t \tau \left(\sum_{n=1}^\infty I_{[| b_n \xi_n | > t]} \right)$$

$$\leq \sup_{t \leq \| \{b_n\} \|_\phi} t \tau \left(\frac{C' \left[\nu\left(\frac{t}{\| \{b_n\} \|_\phi} \right) \right]^{-1}}{P\left(\sup_{n \geq 1} |b_n \xi_n| > t \right)} \right) P\left(\sup_{n \geq 1} |b_n \xi_n| > t \right).$$

Now recall that $\tau(x) = \left[\nu^{-1}\left(\frac{1}{x} \right) \right]^{-1}$ is concave for $x \geq 1$, $\tau(1) = 1$ and

also is regularly varying of index $1/2 < a < 1$. Let

$\alpha = \sup\{x: \tau(x) \geq x\}$. Because τ is regularly varying with index less

than 1, α is finite. One can readily check that for $x \geq 1$ and

$\alpha' \geq \alpha$, $\tau(\alpha' x) \leq \alpha' \tau(x)$. Using this on the right side of the inequality

in (4.109) we get that it is

(4.110) $\leq (C' \vee \alpha) \sup_{t \leq \|\{b_n\}\|_\phi} t\tau\left(\left[\nu\left(\frac{t}{\|\{b_n\}\|_\phi}\right)\right]^{-1}\right)$.

By definition

(4.111) $\tau\left(\left[\nu\left(\frac{1}{x}\right)\right]^{-1}\right) = x, \ \forall x \geq 1$.

Using (4.109), (4.110) and (4.111) we get

(4.112) $\sup_{t \leq \|\{b_n\}\|_\phi} E t \tau\left(\sum_{n=1}^\infty I_{[|b_n \xi_n| > t]}\right) \leq (C' \vee \alpha)\|\{b_n\}\|_\phi$

where the constant $(C' \vee \alpha)$ is independent of the sequence $\{b_n\}_{n=1}^\infty$.
Thus we have established (4.99).

To obtain (4.98) we use Lemma 4.3, in particular the last inequality
in (4.22). This can be evaluated using (4.99) and (4.106). This
completes the proof of Theorem 4.10.

Proof of Theorem 1.2 part II.): By (1.8) only a countable number of
$\{a_\gamma\}_{\gamma \varepsilon A}$ are non-zero. Therefore we can order them and write Y in
(1.18) as

(4.113) $Y(t) = \sum_{n=1}^\infty a_n \xi_n \gamma_n(t), \quad t \ \varepsilon \ K$.

Let $\{\xi_n\}_{n=1}^\infty$ be defined on the probability space (Ω, F, P) and let
$\{\varepsilon_n\}_{n=1}^\infty$ be a Rademacher sequence defined on the probability space
(Ω', F', P'). Let $\omega \ \varepsilon \ \Omega$ and $\omega' \ \varepsilon \ \Omega'$. We will study the series

(4.114) $Y(t; \omega, \omega') = \sum_{n=1}^\infty \varepsilon_n(\omega') a_n \xi_n(\omega) \gamma_n(t), \quad t \ \varepsilon \ K$

defined on the probability space $(\Omega \times \Omega, F \times F', P \times P')$ which is equivalent
to the process in (4.113).

We begin by fixing $\omega \ \varepsilon \ \Omega$ and considering the marginal process
$Y(t; \omega, \cdot)$ as a stochastic process on (Ω', F', P'). Let $\tau(n)$,
$n = 1, 2, \ldots$ be as defined in (4.96). By Lemma 4.1 we have

(4.115) $\| \sum\limits_{n=1}^{\infty} \varepsilon_n a_n \xi_n(\omega)(\Upsilon_n(s) - \Upsilon_n(t)) \|_{\Psi_q} \le K \| \{a_n \xi_n(\omega)(\Upsilon_n(s) - \Upsilon_n(t))\} \|_{\tau,\infty}$

$$\equiv d_\omega(s,t)$$

where

(4.116) $q(n) = n/\tau(n) = n \nu^{-1}(\tfrac{1}{n})$, $n = 1,2,\ldots$.

Extend q as in (4.3) and note that, since $\nu^{-1}(\tfrac{1}{x})$ is regularly varying

as $x \to \infty$ with index greater than -1, we have $\lim \dfrac{H(x)}{x\nu^{-1}(\tfrac{1}{x})} = $ Const. for

H as given in (1.23). It follows from this fact, Lemma 4.2 and (4.116)

that the marginal series $\Upsilon(t;\omega,\cdot)$ has a version with continuous paths if

(4.117) $\int\limits_0^{\hat{d}_\omega} H\big(\log N(K,d_\omega;\varepsilon)\big)d\varepsilon < \infty$

where d_ω is given in (4.11) and $\hat{d}_\omega = \sup\limits_{s,t\in K} d_\omega(s,t)$. The random metric

entropy integral in (4.117) is completely analogous to the one in (4.78).

The same argument used from (4.78) to the end of the proof of Theorem 1.1

part II.) shows that in order to complete this proof we need only show

that

$$E\, d_\omega(s,t) \le C\, d_{\Upsilon,\phi}(s,t), \qquad \forall s,\, t \in K .$$

(This is precisely what is shown in Theorem 4.10, take

$b_n = a_n(\Upsilon_n(s) - \Upsilon_n(t))$. This completes the proof of Theorem 1.2

part II.)

Remark 4.11: Corollary 1.3 is essentially an immediate consequence of

Theorem 1.1. However in proving sufficient conditions in the case where

we take $\eta(u) = \Phi(u)$ we need to check that (1.13) is satisfied, that is,

that both $\eta^{-1}(u)$ and $u/\eta^{-1}(u)$ are concave for $u \ge 1$. One way to see

that this is possible is the following: Begin with

$$\eta^{-1}(u) \equiv \frac{u^{1/p}}{(\log(e+u))^{\beta/p}} , \qquad u \ge u_0'$$

for u_0' sufficiently large and define $\Phi(u)$, $u \geq u_0'$ as the inverse of η^{-1}. It is obvious both that this function Φ satisfies (1.54) and that (1.13) is satisfied at least for $u \geq u_0''$ for some u_0'' sufficiently large. But this is all that matters since $\eta^{-1}(u)$ can always be extended so that the various conditions are satisfied for all $u \geq 1$. This same comment applies (with obvious modifications) in proving the sufficient part of Corollary 1.4 when we take $\nu(n) = \phi(n)$ and need to check that (1.27) is satisfied.

5. PROCESSES FOR WHICH THE LEVY TRANSFORMS OR THE LOGARITHMS OF THE

   CHARACTERISTIC FUNCTIONS ARE REGULARLY VARYING WITH INDEX $1 < p < 2$

The main purpose of this chapter is to prove Theorems 1.7 and 1.8 and to use them to generate many examples of continuous and discontinuous ξ-radial processes and random Fourier series. These examples are used to justify the conjecture about ξ-radial processes given in Chapter I and to show that there is no integral condition involving metric entropy for the random Fourier series except in the stable case. (The 1-stable case is still unresolved.)

Theorem 1.7 and 1.8 give a nice generalization of a classical result of Paley and Zygmund [PZ]. Generalizations of this result have already been given in Theorem 1.6, Chapter VII [MP1] and Lemma 2.3 [MP3]. The results we give here continue the approach used in the latter reference.

We continue to denote by G a locally compact Abelian group with dual group Γ and by K a fixed compact neighborhood of the unit element of G. Following [MP2], Definition 6.1 a subset $\Lambda = \{\gamma_i \mid i \in N\} \subset \Gamma$ is called a topological Sidon set with respect to $K \subset G$ if there exists a constant $C > 0$ such that

(5.1) $\qquad \forall n \in N, \quad \forall \{\alpha_i\} \in \mathbb{C}^N, \quad \sum_{i=1}^{n} |\alpha_i| \leq C \sup_{t \in K} \left| \sum_{i=1}^{n} \alpha_i \gamma_i \right|.$

Now let $\{A_i \mid i \in \mathbb{N}\}$ be disjoint subsets of Γ. We say that $\{A_i\}$ is a topological Sidon partition with respect to $K \subset G$ if all subsets $\{\gamma_i \mid i \in \mathbb{N}\}$ with $\gamma_i \in A_i$ for each $i \in \mathbb{N}$ are topological Sidon sets satisfying (5.1).

The first result of this section deals with series of the type (1.18). To avoid trivialities in what follows we will assume that the random variable ξ defined in (1.17) is not identically zero.

<u>Theorem 5.1</u>:   Consider the random Fourier series defined in (1.17) and

(1.18). Let $\{A_i \mid i \in N\}$ be a topological Sidon partition of Γ with

respect to $K \subset G$ and let A be a countable subset of Γ. Suppose also

that G is a compact group. Then $\forall n$

$$(5.2) \qquad E \parallel \sum_{\gamma \in A} a_\gamma \xi_\gamma \gamma \parallel_\infty \geq C[\sum_{i=1}^n (\sum_{\gamma \in A_i \cap A} |a_\gamma|^2)^{1/2}$$

$$+ \sum_{i=1}^n \parallel \{a_\gamma\}_{\gamma \in A_i \cap A} \parallel_\xi]$$

where for any function $f(t)$, $t \in K$ we define $\parallel f \parallel_\infty = \sup_{t \in K} |f(t)|$ and

$$(5.3) \qquad \alpha_i = \parallel \{a_\gamma\}_{\gamma \in A_i \cap A} \parallel_\xi = \inf\{\lambda > 0 : \sum_{\gamma \in A_i \cap A} P(|a_\gamma \xi| > \lambda) \leq \tfrac{1}{2}\}$$

(Here C is an absolute constant.)

Furthermore, if $\sum_{\gamma \in A} a_\gamma \xi_\gamma \gamma$ converges uniformly a.s. then

$$(5.4) \qquad \sum_{i=1}^\infty \int_{\alpha_i \wedge 1}^1 (\sum_{\gamma \in A_i \cap A} P(|a_\gamma \xi| > u))du < \infty .$$

In general, even if G is not compact we have that if $\sum_{\gamma \in A} a_\gamma \xi_\gamma \gamma$

converges uniformly a.s. the two sums to the right of the inequality in

(5.2) (as well as (5.4)) must be finite.

We will use the following simple Lemma in the proof of Theorem 5.1.

<u>Lemma 5.2</u>: Let $\{\xi_k\}_{k=1}^\infty$ be independent real valued random variables and

$\{a_k\}_{k=1}^\infty \in \mathbb{C}^{\mathbb{N}}$ define

$$(5.5) \qquad \bar\lambda_n = \inf\{\lambda > 0: \sum_{k=1}^n P(|a_k \xi_k| > \lambda) \leq \tfrac{1}{2}\}$$

then for $\forall 0 < p < \infty$ and $\forall n \in \mathbb{N}^+$

(5.6)
$$E\left(\sum_{k=1}^{\infty} |a_k\xi_k|^p\right)^{1/p} \geq \overline{\lambda}_n(1 - e^{-1/2})$$

and

(5.7)
$$E\left(\sum_{k=1}^{\infty} |a_k\xi_k|^p\right)^{1/p} \geq \frac{1}{4} \int_{\overline{\lambda}_n}^{\infty} \left(\sum_{k=1}^{n} P(|a_k\xi_k| > \lambda)\right)d\lambda$$

<u>Proof</u>: We have

(5.8)
$$P\left[\sup_{1\leq k\leq n} |a_k\xi_k| > \lambda\right] = 1 - \prod_{k=1}^{n} \left(1 - P(|a_k\xi_k| > \lambda)\right)$$

$$\geq 1 - \left(\exp - \sum_{k=1}^{n} P(|a_k\xi_k| > \lambda)\right).$$

If $\overline{\lambda}_n = 0$ in (5.6) there is nothing to prove. If $\overline{\lambda}_n > 0$ we have from (5.8) and (5.6) that $\forall \varepsilon > 0$

$$P\left[\sup_{1\leq k\leq n} |a_k\xi_k| > \overline{\lambda}_n - \varepsilon\right] \geq 1 - e^{-1/2}.$$

Therefore

(5.9)
$$E\left(\sum_{k=1}^{\infty} |a_k\xi_k|^p\right)^{1/p} \geq \int_0^{\infty} P\left(\sup_{1\leq k\leq n} |a_k\xi_k| > \lambda\right)d\lambda$$

$$\geq \int_0^{\overline{\lambda}_n-\varepsilon} (1 - e^{-1/2})d\lambda = (\overline{\lambda}_n - \varepsilon)(1 - e^{-1/2}),$$

and since this is true for all $\varepsilon > 0$ we get (5.6).

On the other hand $1-e^{-x} > x/2$ if $0 \leq x \leq 1$. Therefore for $\lambda \geq \overline{\lambda}_n$, by (5.8) we have that

(5.10)
$$P\left(\sup_{1\leq k\leq n} |a_k\xi_k| > \lambda\right) \geq \frac{1}{4} \sum_{k=1}^{n} P(|a_k\xi_k| > \lambda)$$

Substituting this in (5.9) we get (5.7).

<u>Proof of Lemma 5.1:</u>   Let us assume first that G is a compact group.

Without loss of generality we take $K = G$. Let $\{\varepsilon_\gamma\}_{\gamma \in A}$ be a Rademacher

sequence independent from $\{\xi_\gamma\}_{\gamma \in A}$. Let E_ε denote expectation with

respect to $\{\varepsilon_\gamma\}_{\gamma \in A}$ and E_ξ denote expectation with respect to $\{\xi_\gamma\}_{\gamma \in A}$.

By Theorem 1.4, Chapter I, [MP1] and (2.11) of Lemma 2.3 [MP3] we have

$$(5.11) \qquad E_\varepsilon \| \sum_{\gamma \in A} a_\gamma \varepsilon_\gamma \xi_\gamma \gamma \|_\infty \geq C \sum_{i=1}^n (\sum_{\gamma \in A_i \cap A} |a_\gamma \xi_\gamma|^2)^{1/2} \, ,$$

where C is a constant independent of $\{a_\gamma\}$ and A. By (5.6) we have

$$(5.12) \qquad E\| \sum_{\gamma \in A} a_\gamma \xi_\gamma \gamma \|_\infty = E_\xi E_\varepsilon \| \sum_{\gamma \in A} a_\gamma \varepsilon_\gamma \xi_\gamma \gamma \|_\infty$$

$$\geq C \sum_{i=1}^n E(\sum_{\gamma \in A_i \cap A} |a_\gamma \xi_\gamma|^2)^{1/2}$$

$$\geq C(1 - e^{-1/2}) \sum_{i=1}^n \| \{a_\gamma\}_{\gamma \in A_i \cap A} \|_\xi \, .$$

Also, by Theorem 5.3 [JM1], since $\{\xi_\gamma\}$ are i.i.d. and not identically

zero we have

$$(5.13) \qquad E\| \sum_{\gamma \in A} a_\gamma \xi_\gamma \gamma \|_\infty \geq k \, E\| \sum_{\gamma \in A} a_\gamma \varepsilon_\gamma \gamma \|_\infty$$

for some constant $k > 0$. Therefore by (5.11) and (5.13) we get

$$(5.14) \qquad E\| \sum_{\gamma \in A} a_\gamma \xi_\gamma \gamma \|_\infty \geq Ck \sum_{i=1}^n (\sum_{\gamma \in A_i \cap A} |a_\gamma|^2)^{1/2} \, .$$

Combining (5.12) and (5.14) we get (5.2) in the case when G is compact

and $K = G$.

Even if G is not compact, $\sum_{\gamma \in A} a_\gamma \varepsilon_\gamma \xi_\gamma \gamma$ converging uniformly a.s

implies by Theorem 1.1, Chapter I, [MP1] and Lemma 2.3 [MP3] that

$$(5.15) \qquad \sum_{i=1}^\infty (\sum_{\gamma \in A_i \cap A} |a_\gamma \xi_\gamma|^2)^{1/2} < \infty \quad \text{a.s.}$$

Obviously (5.15) implies that

$$\sum_{i=1}^{\infty} \left(\sup_{\gamma \in A_i \cap A} |a_\gamma \xi_\gamma| \right) < \infty \quad \text{a.s.}$$

and this, by the Three Series Theorem implies that

$$(5.16) \qquad \sum_{i=1}^{\infty} E\left(\sup_{\gamma \in A_i \cap A} |a_\gamma \xi_\gamma| \, I_{\left[\sup_{\gamma \in A_i \cap A} |a_\gamma \xi_\gamma| \leq 1 \right]} \right) < \infty .$$

Following the proof of Lemma 5.2 we see that

$$(5.17) \quad E\left(\sup_{\gamma \in A_i \cap A} |a_\gamma \xi_\gamma| \, I_{\left[\sup_{\gamma \in A_i \cap A} |a_\gamma \xi_\gamma| \leq 1 \right]} \right) \geq \int_0^1 P\left(\sup_{\gamma \in A_i \cap A} |a_\gamma \xi_\gamma| > \lambda \right) d\lambda$$

$$\geq \left(\| \{a_\gamma\}_{\gamma \in A_i \cap A} \|_\xi \wedge 1 \right) (1 - e^{-1/2}).$$

By (5.16) and (5.17) we see that

$$\sum_{i=1}^{\infty} \left(\| \{a_\gamma\}_{\gamma \in A_i \cap A} \|_\xi \wedge 1 \right) < \infty .$$

But since $\| \{a_\gamma\}_{\gamma \in A_i \cap A} \|_\xi \to 0$ as $i \to \infty$ we see that the second series to the right of the inequality in (5.2) must converge. The convergence of the first term to the right of the inequality in (5.2) follows from Theorem 2.2, [MP3].

Also, using (5.17) and (5.10) we see that the left side of (5.17)

$$\geq \int_{\alpha_i \wedge 1}^1 \sum_{\gamma \in A_i \cap A} P(|a_\gamma \xi_\gamma| > \lambda) d\lambda$$

Thus we get (5.4). This completes the proof of lemma 5.1.

Although (5.2) has a nice form and is useful when $P(|\xi| > \lambda)$ is regularly varying at infinity with index $\alpha < -1$ the first inequality in (5.12) is strictly sharper than (5.2). This can be readily seen in the case when ξ is a 1-stable random variable. In general, if $P(|\xi| > \lambda)$ is regularly varying of index -1 we use (5.4). (When $P(|\xi| > \lambda)$ is regularly varying of index $\alpha < -1$, (5.2) and (5.4) are equivalent.)

Remark 5.3. The choice of 1/2 in (5.3) rather than 1 was to avoid having $\lambda = 0$ when $\{a_\gamma\}_{\gamma \in A_i \cap A}$ contains only one non-zero element. If $P(|\xi| > \lambda)$ is regularly varying at infinity there is lattitude for our definition of $\| \ \|_\xi$. To be more explicit let us note the following: For $0 < c \le 1/2$ define, for I some subset of the integers,

$$(5.18) \qquad \lambda_c = \inf \left\{ \lambda > 0 : \sum_{k \in I} P(|a_k \xi| > \lambda) \le c. \right\}$$

and assume that $P(|\xi| > \lambda)$ is regularly varying at infinity with index $-2 < p < 0$. Then for $0 < c \le c' \le 1/2$

$$(5.19) \qquad k\lambda_c \le \lambda_{c'} \le \lambda_c$$

where k depends on c but not on $\{a_k\}_{k \in I}$. The right side of (5.19) is obvious. To obtain the left side we need to show that for fixed c and c'

$$(5.20) \qquad \sum_{k \in I} P(|a_k \xi| > k^{-1} \lambda_{c'}) \le c.$$

Since λ_c is homogeneous we can assume that $P(|\xi| > 1) \ge \frac{1}{2}$ this implies that for $0 < c \le 1/2$, $\lambda_c \ge \lambda_{1/2} \ge \sup_{k \in I} |a_k|$. By the representation for regularly varying functions, $P(|\xi| > \lambda) = \lambda^p L(\lambda)$ where L is slowly varying at infinity. Thus

$$P(|a_k \xi| > \delta \lambda_{c'}) = \left| \frac{a_k}{\lambda_{c'} \delta} \right|^p L\left(\frac{\delta \lambda_{c'}}{|a_k|} \right)$$

$$= P(|a_k \xi| > \lambda_{c'}) \, \delta^{-p} L\left(\frac{\delta \lambda_{c'}}{|a_k|} \right) L^{-1}\left(\frac{\lambda_{c'}}{|a_k|} \right).$$

Since $\lambda_{c'} / |a_k| \ge 1$, we see by Chapter VIII (8.5) [F] that

$$\lim_{\delta \to \infty} \delta^{-p} L\left(\frac{\delta \lambda_{c'}}{|a_k|} \right) / L\left(\frac{\lambda_{c'}}{|a_k|} \right) = 0,$$

and so we can find a value of δ such that

(5.21) $$P(|a_k\xi| > \delta\lambda_{c'}) \leq \frac{c}{c'} P(|a_k\xi| > \lambda_{c'}) \ .$$

Summing (5.21) over $k \in I$ and using (5.18) we get (5.20) with
$k = \delta^{-1}$. Thus we have obtained (5.19) we will need this in the proof of
Theorem 1.8.

We next give a lower bound for ξ-radial processes. It follows from
Lemma 2.3 and Corollary 2.4 that for every Levy measure τ and
probability measure m on Γ we can define a real valued strongly
stationary ξ-radial process by the series

(5.22) $$\sum_{j=1}^{\infty} \varepsilon_j F^{-1}(\Gamma_j) Y_j(t), \quad t \in K$$

and a complex valued strongly stationary ξ-radial process by the series

(5.23) $$\sum_{j=1}^{\infty} F^{-1}(\Gamma_j) e^{i\theta_j} Y_j(t), \quad t \in K \ .$$

(See Lemma 2.3 for the precise definitions of the terms in (5.22) and
(5.23).) Furthermore, it follows from Theorem 4.9, Chapter II, [MP1] that
$E\|X(t)\|_\infty < \infty$ if and only if $E\|Z(t)\|_\infty < \infty$ and that the series
representing $X(t)$ converges uniformly a.s. if and only if the series
representing $Z(t)$ converges uniformly a.s. Therefore, in the next
theorem, we only have to give results for the series in (5.22).

<u>Theorem 5.3.</u> Let $\{A_i \mid i \in N\}$ be a typological Sidon partition of Γ
with respect to $K \subset G$ and let A be a countable subset of Γ. Let

$$m_i = m\{\gamma \in \Gamma \mid \gamma \in A_i\} \ .$$

Suppose also that G is a compact group. Then for all integers $j_0 \geq 1$

$$(5.24) \quad E\| \sum_{j=j_0}^{\infty} \varepsilon_j F^{-1}(\Gamma_j)Y_j \|_\infty \geq C \inf_{j \geq j_0} E\left(\frac{F^{-1}(\Gamma_j)}{F^{-1}(j)}\right) \sum_{i=1}^{\infty} m_i \left(\sum_{j=j_0}^{[m_i^{-1}]+1} F^{-1}(j) \right)$$

where C is an absolute constant and we define $0 \cdot \infty = 0$.

In general even if G is not compact, if the series in (5.22) converges uniformly a.s. then the series on the right in (5.24) converges for some (equivalently all) $j_0 \geq 1$.

Proof: Let us begin with the case when G is a compact group. Recall that $\{\varepsilon_j\}_{j=1}^{\infty}$, $\{\Gamma_j\}_{j=1}^{\infty}$ and $\{Y_j\}_{j=1}^{\infty}$ are all independent of each other. Let E_ε and E_Γ denote expectations with respect to $\{\varepsilon_j\}_{j=1}^{\infty}$ and $\{\Gamma_j\}_{j=1}^{\infty}$. It follows from convexity and (4.8) of Theorem 4.9, Chapter II, [MP1] that

$$(5.25) \quad E_\varepsilon E_\Gamma \| \sum_{j=j_0}^{\infty} \varepsilon_j F^{-1}(\Gamma_j)Y_j \|_\infty \geq \inf_{j \geq j_0} E\left(\frac{F^{-1}(\Gamma_j)}{F^{-1}(j)}\right) E_\varepsilon \| \sum_{j=j_0}^{\infty} \varepsilon_j F^{-1}(j)Y_j \|_\infty.$$

By Theorem 1.4, Chapter I, [MP1] and (2.11) of Lemma 2.3, [MP3] this last term

$$(5.26) \quad \geq \inf_{j \geq j_0} E\left[\left(\frac{F^{-1}(\Gamma_j)}{F^{-1}(j)}\right) \sum_{i=1}^{\infty} \left(\sum_{j=j_0}^{\infty} (F^{-1}(j))^2 I_{[Y_j \in A_i]} \right)^{1/2} \right].$$

Therefore

$$(5.27) \quad E\| \sum_{j=j_0}^{\infty} \varepsilon_j F^{-1}(\Gamma_j)Y_j \|_\infty \geq \inf_{j \geq j_0} E\left(\frac{F^{-1}(\Gamma_j)}{F^{-1}(j)}\right) \sum_{i=1}^{\infty} E(\sup_{j \geq j_0} F^{-1}(j)I_{[Y_j \in A_i]}).$$

Clearly the probability, that the smallest value of j for which $Y_j \in A_i$ is k, is $(1-m_i)^{k-1}m_i$ since $P(Y_j \in A_i) = m_i$. Thus

$$(5.28) \quad E\sup_{j \geq j_0} F^{-1}(j)I_{[Y_j \in A_i]} = \sum_{j=j_0}^{\infty} (1-m_i)^{j-1}m_i F^{-1}(j).$$

Suppose that $m_{i'} > 1/2$ for some i'. It is easy to see that the left side of (5.27)

$$(5.29) \qquad \geq \frac{1}{2} \inf_{j \geq j_0} E \left(\frac{F^{-1}(\Gamma_j)}{F^{-1}(j)} \right) F^{-1}(j_0)$$

$$\geq \frac{1}{4} \inf_{j \geq j_0} E \left(\frac{F^{-1}(\Gamma_j)}{F^{-1}(j)} \right) m_i, \left(\sum_{j=j_0}^{[m_i^{-1}]+1} F^{-1}(j) \right).$$

Therefore without loss of generality we can assume that $m_i \leq 1/2$. This implies that the last term in (5.28)

$$(5.30) \qquad \geq m_i \sum_{j=j_0}^{[m_i^{-1}]+1} (1 - m_i)^{j-1} F^{-1}(j)$$

$$\geq e^{-2} m_i \sum_{j=j_0}^{[m_i^{-1}]+1} F^{-1}(j).$$

Combining (5.25) - (5.28) and (5.30) we get (5.24).

Suppose now that G is not necessarily compact. By the strong law of large numbers we know that the series in (5.22) converges uniformly a.s. if and only if

$$E \| \sum_{j=1}^{\infty} \varepsilon_j F^{-1}(j) Y_j \|_{\infty} < \infty.$$

We can proceed, exactly as in the first part of this proof, to show that the sum on the right in (5.24) converges with $j_0 = 1$. This completes the proof of this Theorem.

To see the significance of j_0 in (5.24) note that for certain Levy measures τ, $EF^{-1}(\Gamma_1) = \infty$. This is the case for example for the Levy measures corresponding to p-stable processes, for $p \leq 1$. Nevertheless, in all these cases, $EF^{-1}(\Gamma_j) < \infty$ for j sufficiently large. Thus by looking at the series on the left in (5.24) starting from the j_0-th term we can restrict our consideration to series which converge uniformly a.s. if and only if the expected value of their sup-norm is finite.

We now proceed to the proofs of Theorem 1.7 and 1.8.

Proof of Theorem 1.7: Since Ψ is regularly varying at infinity we have by (4.71a) that there exist constants c, C and $u_0 > 0$ such that

$$(5.31) \qquad c\Psi\left(\frac{1}{u}\right) < \tau\left[\frac{1}{u}, \infty\right) < C\Psi\left(\frac{1}{u}\right), \quad 0 < u \leq u_0,$$

where τ is the Levy measure associated with Ψ. By (5.31)

$$\tau\left[(\Psi^{-1}(t/c))^{-1}, \infty\right) > t$$

consequently the function F^{-1} defined in Lemma 2.1 satisfies

$$(5.32) \qquad F^{-1}(t) \geq (\Psi^{-1}(t/c))^{-1} \geq \frac{c^{1/p}}{2}(\Psi^{-1}(t))^{-1}$$

for t sufficiently large.

The ξ-radial process that this Theorem refers to can be represented as

$$X(t) = \sum_{j=1}^{\infty} F^{-1}(\Gamma_j)\tilde{g}_j Y_j(t), \quad t \in [0, 2\pi].$$

By (5.24) and (5.31) we see that

$$(5.33) \qquad E\|X\|_{\infty} \geq cE\left\|\sum_{j=1}^{\infty} \varepsilon_j F^{-1}(\Gamma_j) Y_j\right\|_{\infty}$$

$$\geq c' \sum_{i \in I} m_i \left(\sum_{j=1}^{[m_i^{-1}]} (\Psi^{-1}(j))^{-1}\right)$$

where $c, c' > 0$ are constants independent of $\{m_i\}_{i=1}^{\infty}$ and where, in order to fullfill the criterion of topological Sidon partition, we let I denote either the even integers or the odd integers. (We can take $j_0 = 1$ in (5.24) because $EF^{-1}(\Gamma_j) < \infty$ in these cases.) Because Ψ^{-1} is regularly varying at infinity with index less than one

$$(5.34) \qquad \sum_{j=1}^{[m_i^{-1}]+1} (\Psi^{-1}(j))^{-1} \geq k(m_i \ \Psi^{-1}(\tfrac{1}{m_i}))^{-1},$$

for some constant $k > 0$. Substituting this in (5.33) we get the necessary condition of this theorem.

We now obtain conditions for continuity. A ξ-radial process as defined by (1.5) for which the function $\Psi(|u|)$ satisfies the conditions of the hypothesis can be represented by the series

$$(5.35) \qquad \sum_{j=1}^{\infty} F^{-1}(\Gamma_j) \ e^{i\theta_j} \ Y_j(t), \qquad t \ \varepsilon \ [0, 2\pi]$$

where

$$\forall j, \qquad P[Y_j(t) = e^{ikt}] = m\{e^{ikt}\} \equiv a_k .$$

Furthermore we know from Corollary 2.4 that there exists a real valued infinitely divisible random variable η such that

$$E \ \exp \ i\lambda\eta = \exp - \psi(|\lambda|)$$

and such that

$$\Psi(|\lambda|) = E_\theta \ \psi(|\lambda \cos \theta|) .$$

Let τ be the Levy measure associated with η and let F^{-1} be defined in terms of τ in the usual way. Consider

$$(5.36) \qquad \tilde{\xi}_k = \sum_{k=1}^{\infty} F^{-1}(\Gamma_j) e^{i\theta_j} \delta_j$$

where $\{\delta_j\}_{j=1}^{\infty}$ are i.i.d. with $P(\delta_1 = 1) = a_k$ and $P(\delta_1 = 0) = 1 - a_k$. Since

$$Ee^{i \ Re(\bar{z} \ \tilde{\xi}_k)} = \exp - E_\theta E_\delta \psi(|\bar{z}(\cos \theta)\delta|)$$

$$= \exp - a_k E_\theta \psi(|z \cos \theta|) = \exp - a_k \Psi(|z|) ,$$

one sees that the series

(5.37)
$$\tilde{H}(t) = \sum_{k=0}^{\infty} \tilde{\xi}_k e^{ikt}, \quad t \in [0, 2\pi] \ ,$$

where the $\{\tilde{\xi}_k\}_{k=0}^{\infty}$ are independent, has the same characteristic functional as the series in (5.35). Also, clearly, we have that $\text{Re } \tilde{\xi}_k = \text{Im } \tilde{\xi}_k$ and, by (2.4), $\xi_k \equiv \text{Re } \tilde{\xi}_k$ satisfies

(5.38)
$$E \ e^{i\lambda \xi_k} = \exp -a_k \Psi(|\lambda|), \quad -\infty < \lambda < \infty \ .$$

Thus we see that to show that the ξ-radial process under consideration has continuous sample paths it is enough to show that the series

(5.39)
$$H(t) = \sum_{k=0}^{\infty} \xi_k e^{ikt}, \quad t \in [0, 2\pi]$$

has continuous sample paths when $\{\xi_k\}$ is independent and satisfies (5.38).

We shall show shortly that

(5.40)
$$\sum_{k=1}^{\infty} P(|\xi_k| > 1) < \infty$$

so in what follows we can replace $\{\xi_k\}_{k=1}^{\infty}$ by

$$\xi_k' = \xi_k I_{[|\xi_k| \leq 1]}, \quad \forall k \geq 1 \ .$$

In order to show that the series in (5.39) converges uniformly a.s. it is enough to show that for some positive sequence $\{b_k\}$ with $\lim_{k \to \infty} b_k = 0$ both the series

(5.41)
$$H_1(t) = \sum_{k=k_0}^{\infty} \xi_k' \ I_{[|\xi_k'| > b_k]} e^{ikt}, \quad t \geq [0, 2\pi]$$

and

(5.42)
$$H_2(t) = \sum_{k=k_0}^{\infty} \xi_k I_{[|\xi_k| \leq b_k]} e^{ikt}, \quad t \in [0, 2\pi]$$

converge uniformly a.s. Following the approach of [CL] we will obtain the uniform convergence a.s. of $\{H_1(t), \ t \in [0, 2\pi]\}$ by showing that

$$(5.43) \qquad \sum_{k=k_0}^{\infty} E|\xi_k'| \, I_{[\,|\xi_k'|>b_k]} < \infty$$

and that of $H_2(t)$ by showing that

$$(5.44) \qquad \sum_{n=n_0}^{\infty} \frac{\left(\sum_{k=2^n}^{\infty} E|\xi_k|^2 I_{[\xi_k \le b_k]} \right)^{1/2}}{n^{1/2}} < \infty \; .$$

If (5.43) holds $\{H_1(t), \; t \in [0,2\pi]\}$ converges absolutely. That (5.44) is sufficient for the continuity of $\{H_2(t), \; t \in [0,2\pi]\}$ follows from the classical result of Salem and Zygmund, see (1.5) and (1.9), Chapter I, and Lemma 1.1 Chapter VII, [MP1] for more details.)

By a well known inequality, see pg. 196 [L] we have, $\forall \, k \ge 0$,

$$(5.45) \qquad P(|\xi_k| > u) \le 7u \int_0^{1/u} (1 - \exp(-a_k \Psi(|\lambda|)))d\lambda$$

$$\le 7a_k u \int_0^{1/u} \Psi(|\lambda|)d\lambda$$

since $1 - e^{-x} \le x$. By hypothesis $\Psi(|\lambda|)$ is regularly varying at infinity with index greater than 1. Therefore there exists a constant c independent of the sequence $\{a_k\}_{k=1}^{\infty}$ such that

$$(5.46) \qquad P(|\xi_k| > u) \le c a_k \, \Psi(|\tfrac{1}{u}|), \quad u \le 1 \; .$$

Note that (5.40) follows from (5.46).

We have, by (5.46)

$$E|\xi_k'| \, I_{[\,|\xi_k'|>b_k]} \le \int_{b_k}^{\infty} P(|\xi_k'| > u) \, du$$

$$\le c a_k \int_{b_k}^{\infty} \Psi(|\tfrac{1}{u}|) du$$

$$= c a_k \int_1^{1/b_k} \frac{\Psi(|s|)}{s^2} \, ds$$

$$\le c' a_k b_k \Psi(|\tfrac{1}{b_k}|)$$

where, at the last step, we use the fact that $\Psi(|u|)$ is regularly varying with index greater than 1. Therefore in order to obtain (5.43) we must show that

$$(5.47) \qquad \sum_{n=n_0}^{\infty} \sum_{k=2^n}^{2^{n+1}-1} a_k b_k \Psi\left(\left|\frac{1}{b_k}\right|\right) < \infty \ .$$

To obtain (5.44) we note that by (5.46)

$$E|\xi_k|^2 \ I_{[\xi_k \leq b_k]} \leq 2 \int_0^{b_k} u P(|\xi_k| > u) \ du$$

$$\leq 2ca_k \int_0^{b_k} u\Psi\left(\left|\frac{1}{u}\right|\right) du$$

$$= 2ca_k \int_{1/b_k}^{\infty} \frac{\Psi(|s|)}{s^3} \ ds$$

$$< c'a_k b_k^2 \ \Psi\left(\left|\frac{1}{b_k}\right|\right)$$

where, at the last stage, we use the fact that $\Psi(|s|)$ is regularly varying of index less than 2. Therefore for u sufficiently large

$$(5.48) \qquad \sum_{k=2^n}^{\infty} E|\xi_k|^2 \ I_{[|\xi_k| \leq b_k]} \leq c' \sum_{k=2^n}^{\infty} b_k^2 a_k \ \Psi\left(\left|\frac{1}{b_k}\right|\right) \ .$$

In (5.47) and (5.48) set

$$(5.49) \qquad b_k = (\Psi^{-1}(\frac{1}{m_n}))^{-1}, \quad 2^n \leq k < 2^{n+1}$$

and note that

$$m_n = \sum_{k=2^n}^{2^{n+1}-1} a_k \ .$$

We see immediately that (1.70) implies (5.47) is finite. By (5.48) and (5.49) we see that (5.44) holds if

$$(5.50) \qquad \sum_{n=n_0}^{\infty} \frac{\left(\sum_{k=n}^{\infty} (\Psi^{-1}(\frac{1}{m_n}))^{-2} \right)^{1/2}}{n^{1/2}} < \infty.$$

It follows from Chapter IV Lemma 2.2 [JM], since $\Psi^{-1}(\frac{1}{m_n})$ is non-decreasing, that (1.70) is equivalent to (5.50). This completes the proof of Theorem 1.7.

There is actually no loss in generality in assuming that $\Psi(u)$ is strictly increasing for u sufficiently large in Theorem 1.7 so that $\Psi^{-1}(u)$ is well defined. By (5.31), $\Psi(u)$ is comparable to the increasing function $\tau[\frac{1}{u},\infty)$ for u sufficiently large. If $\tau[\frac{1}{u},\infty)$ is not continuous it can be modified to be continuous in such a way that the ξ-radial process corresponding to the modified τ, call it $\tilde{\tau}$, has continuous paths if and only if the ξ-radial process corresponding to the original τ does. (This is because continuity only depends on $F^{-1}(j)$ for integers $j \geq j_0$ for some j_0 sufficiently large.) Let $\tilde{\Psi}$ be the Levy transform of $\tilde{\tau}$. Apply Theorem 1.7 to the ξ-radial process determined by $\tilde{\Psi}$. Since $\tilde{\Psi}$ is comparable to $\tilde{\tau}[\frac{1}{u},\infty)$ use $\tilde{\tau}^{-1}[\frac{1}{u},\infty)$ in (1.70). The convergence or divergence of this sum will determine the continuity of the original process.

Let us also note that the necessary condition in Theorem 1.7 is also valid in the context of topological Sidon partitions considered in Theorem 3.

Proof of Theorem 1.8: Since $E|\xi_1| < \infty$ it follows from the Ito-Nisio Theorem and a result of Hoffman-Jorgensen, (see e.g. [HJ] or Theorem 3.3, [JM1]) that if the series $Y(t)$ in (1.71) has bounded paths then $E \sup_{t \in G} |Y(t)| < \infty$. Thus to prove unboundedness a.s. we need only show that the right side of (5.2), with $u = \infty$, is infinite. Since ψ is regularly varying at zero we have that

(5.51) $$P\left(\left| \xi \right| > u \right) > c\psi\left(\tfrac{1}{u}\right), \quad u \geq 1$$

for some constant $0 < c \leq 1/2$.

Let $\lambda_n = \| \{a_k\}_{k=2^n}^{2^{n+1}+1} \|_\psi$ as defined in (1.72). By (5.51)

(5.52) $$\sum_{k=2^n}^{2^{n+1}-1} P\left(\left| a_k \xi \right| > \lambda_n \right) > c \sum_{k=2^n}^{2^{n+1}-1} \psi\left(\frac{|a_k|}{\lambda_n}\right) = c$$

since without loss of generality we can take $\sup_k |a_k| \leq 1$. Define

$$\lambda_{n,c} = \inf \{ \lambda > 0: \sum_{k=2^n}^{2^{n+1}-1} P\left(\left| a_k \xi \right| > \lambda \right) \leq c \} .$$

By (5.52) we have that $\lambda_{n,c} \geq \lambda_n$ and by (5.19) that $\lambda_{n,1/2} \geq k\lambda_n$. Therefore if $\sum \lambda_n = \infty$ then for at least one of the values $i = 1$ or 2

(5.53) $$\sum_{n \in I_i} \lambda_{n,1/2} = \infty$$

where I_1 (I_2) represents the positive odd (even) integers. Let us take $\{A_m\}_{m=1}^\infty = \{e^{ikt}, \ 2^{2m} \leq k < 2^{2m+1}\}_{m=1}^\infty$ or $\{A_m^1\}_{m=1}^\infty = \{e^{ikt}, \ 2^{2m-1} \leq k < 2^{2m}\}_{m=1}^\infty$. These are both topological Sidon partitions of Γ. We see from (5.53) that for at least one of these choices the right side of (5.2), with $n = \infty$, is infinite. This establishes the necessity of (1.73).

In proving continuity conditions let us first notice that these results depend on the tail of the distribution of ξ. That is because, by the contraction principle, we have (see e.g. Theorem 4.9, Chapter II [MP1])

(5.54) $$E\| \sum_{k=1}^\infty a_k \xi_k I_{[\xi_k \leq M]} \gamma_k \|_\infty \ \leq \ 2ME\| \sum_{k=1}^\infty a_k \varepsilon_k \gamma_k \|_\infty$$

where $\{\varepsilon_k\}_{k=1}^{\infty}$ is a Rademacher sequence. As we remarked prior to Theorem 1.7 the series on the right in (5.54) converges a.s. and the expection is finite if and only if $\sum_n (\sum_{k=2^n}^{2^{n+1}-1} |a_k|^2)^{1/2} < \infty$, i.e. (1.73) with $\psi(x) = x^2$. Since the functions ψ that concern us are regularly varying with index less than 2 we will assume that $P(|\xi| \geq 1) = 1$. Since ψ is regularly varying we have that

$$(5.55) \qquad P(|\xi| > u) \leq C \psi \left(\frac{1}{u}\right), \quad u \geq 1 ,$$

for some constant C. We now follow the method used in the proof of Theorem 1.7. We need to show that

$$(5.56) \qquad \sum_{n=n_0}^{\infty} \sum_{k=2^n}^{2^{n+1}-1} a_k E|\xi| I_{[|\xi|>b_k]} < \infty$$

and

$$(5.57) \qquad \sum_{n=n_0}^{\infty} \frac{\left(\sum_{k=2^n}^{\infty} a_k^2 E|\xi|^2 I_{[|\xi|\leq b_k]}\right)^{1/2}}{n^{1/2}} < \infty ,$$

for some sequence $\{b_k\}_{k=1}^{\infty}$ and $n \geq n_0$ sufficiently large. By (5.55) we have

$$(5.58) \qquad E|\xi| I_{[|\xi|>b_k]} \leq \int_{b_k}^{\infty} P[|\xi| > u] du$$

$$\leq C' b_k \psi\left(\frac{1}{b_k}\right) .$$

Similarly

$$E|\xi|^2 I_{[|\xi|\leq b_k]} \leq 2 \int_0^{b_k} u P(|\xi| > u) du$$

$$\leq C' b_k^2 \psi\left(\frac{1}{b_k}\right) .$$

Therefore

$$(5.59) \qquad \sum_{k=n}^{\infty} a_k^2 E |\xi|^2 I_{[|\xi| \leq b_k]} \leq C' \sum_{k=n}^{\infty} a_k^2 b_k^2 \psi\left(\frac{1}{b_k}\right) .$$

We take

$$b_k = (a_k^{-1}) \, \|\{a_k\}_{k=2^n}^{2^{n+1}-1}\|_\psi, \quad 2^n \leq k < 2^{n+1}-1 .$$

Using this in (5.58) and recalling (1.72) we see that (5.56) is satisfied. Using this in (5.59) we see that (5.57) is equivalent to

$$(5.60) \qquad \sum_{n=n_0}^{\infty} \frac{\left(\sum\limits_{k=n}^{\infty} \|\{a_k\}_{k=2^n}^{2^{n+1}-1}\|_\psi^2\right)^{1/2}}{n^{1/2}} < \infty .$$

As we remark following (5.50) the fact that $\|\{a_k\}_{2^n}^{2^{n+1}-1}\|_\psi$ is non-increasing implies that (5.60) holds if and only if (1.73) holds. This completes the proof of Theorem 1.8.

The following Corollaries of Theorems 1.7 and 1.8 are immediate.

<u>Corollary 5.4</u>: Let $G = [0,2\pi]$ so that $\Gamma = \{e^{ikt}, k \in N\}$. Let $\{X(t)\}_{t \in G}$ be a $\mathbb{C}^G$ valued stochastic process with characteristic functional given by (1.5) with $\Psi(u) \sim u^p (\log u)^\beta$, $1 < p < 2$, for u sufficiently large. Let the measure m in (1.5) satisfy

$$(5.61) \qquad m\{e^{ikt}\} = C_{p,\beta} [k(\log k)^p (\log \log k)^{p+\beta+\epsilon}]^{-1} \equiv a_k$$

for $k \geq 10$, $m\{e^{ikt}\} = C_{p,\beta}$, $0 < k < 10$, where $C_{p,\beta}$ is such that $|m| = 1$. Then if $\epsilon > 0$ in (5.61), $\{X(t)\}_{t \in G}$ has a version with continuous sample paths. However, if $\epsilon \leq 0$, $\sup\limits_{t \in G} |X(t)|$ is unbounded a.s.

Let the measure m in (1.5) satisfy

$$(5.62) \qquad m(e^{i2^k t}) = c_{p,\beta} [k^p (\log k)^{p+\beta+\epsilon}]^{-1}$$

for $k \geq 3$, $m(e^{i2^k t}) = c_{p,\beta}$, $0 \leq k < 3$, where $c_{p,\beta}$ is such that $|m| = 1$. Then if $\varepsilon > 0$ in (5.62), $\{X(t)\}_{t\varepsilon G}$ has a version with continuous sample paths. However if $\varepsilon \leq 0$, $\sup\limits_{t\varepsilon G} |X(t)|$ is unbounded a.s.

<u>Corollary 5.5</u>: Let $G = [0,2\pi]$ so that $\Gamma = \{e^{ikt}, k \varepsilon N\}$. Let $\psi(\lambda) \sim \lambda^p(\log 1/\lambda)^\beta$, $1 < p < 2$, for $\lambda > 0$ sufficiently close to zero; let ξ be as given in (1.17) and consider

(5.63) $$Y(t) = \sum_{k=0}^{\infty} a_k \xi_k e^{ikt}, \quad t \varepsilon G$$

where $\{\xi_k\}_{k=0}^{\infty}$ are i.i.d copies of ξ. Let

(5.64) $$a_k = [k^{1/p}(\log k)^{1+\beta/p} (\log \log k)^{1+\varepsilon}]^{-1}$$

$k \geq 10$, otherwise let $a_k = 1$, $k = 1,\ldots,9$. Then if $\varepsilon > 0$ in (5.64) the series in (5.63) converges uniformly a.s. whereas if $\varepsilon \leq 0$, $\sup\limits_{t\varepsilon G} |Y(t)|$ is unbounded a.s.

The examples in Corollaries 5.4 and 5.5 are specific cases of the processes considered in Corollaries 1.5 and 1.6 respectively. Our interest in these examples is to use them to test the integral conditions given in (1.56), (1.59), (1.60), (1.63) and (1.65). What we will show first is that for Φ as given in (1.54) both

(5.65) $$\int_0^{\infty} \big(\log N([0,2\pi], d_{X,\Phi}; \varepsilon)\big)^{1/q} d\varepsilon < \infty$$

and

(5.66) $$J(H_\psi, d_{X,p}) = \int_0^{\infty} H_\psi (\log N[0,2\pi], d_{X,p}; \varepsilon))d\varepsilon < \infty$$

(where $\Psi \sim \Phi$ for x large by hypothesis), for those processes we

considered in Corollary 5.4 which are continuous but that neither (5.65)
nor (5.66) holds when applied to those processes considered in Corollary
5.4 which are unbounded a.s. Here Φ is a convex function equivalent to
Ψ as infinity. At the same time we see that the integral condition in
(1.56), is not a necessary and sufficient condition for the continuity of
the processes considered in Corollary 5.4.

The connection between (5.65) and the processes considered in
Corollary 5.4 is through the metric

$$d_{X,\Phi}(u,0) = \inf\{c > 0 : \sum_{k=0}^{\infty} a_k \Phi(|\frac{2 \sin ku}{c}|) \leq 1\}$$

where $\{a_k\}_{k=0}^{\infty}$ is given in (5.61). Since $d_{X,\Phi}$ is translation invariant
on $[0,2\pi]$ the integral in (5.65) is finite if and only if

(5.67) $\int_0^{\infty} (\log N([-\delta,\delta], d_{X,\Phi};\varepsilon))^{1/q} d\varepsilon < \infty$

for some $\delta > 0$. For technical reasons involved in the estimation of
$d_{X,\Phi}$ it will be more convenient to consider the condition in (5.67)
rather than the one in (5.65).

We will first show that for, $\frac{1}{n+1} < u \leq \frac{1}{n}$, and all $n \geq n_0$
sufficiently large

(5.68) $d_{X,\Phi}(u,0) \leq \alpha[(\log n)^{1/q}(\log \log n)^{1+\varepsilon/p}]^{-1} \equiv \alpha_n$

where α is some constant independent of $n \geq n_0$. In order to obtain
(5.68) it is necessary to define $\Phi(\lambda)$ for all values of λ. However, as
is well known, (see e.g. [KR]), given two convex functions Φ_1 and Φ_2
both satisfying (1.54) there exists a constant k such that for any
random variable X

$$k^{-1}d_{X,\Phi_2}(u,0) \leq d_{X,\Phi_1}(u,0) \leq kd_{X,\Phi_2}(u,0) .$$

Therefore we are free to choose any convex function Φ that satisfies (1.54). For simplicity we choose Φ to satisfy

$$\Phi(u) = u^2 \ , \quad 0 \leq u \leq 1$$

and

$$\Phi(u) = u^p (\log u)^\beta \ , \quad u \geq u_0' > 0$$

for some u_0' sufficiently large. We have, for $\frac{1}{n+1} < u \leq \frac{1}{n}$, for $n \geq n_0$ sufficiently large that

$$(5.69) \qquad E\Phi\left(\frac{|Y(u) - 1|}{\alpha_n}\right) \leq \sum_{k=1}^{n} a_k \Phi\left(\frac{2k}{\alpha_n n}\right) + \sum_{k=n+1}^{\infty} a_k \Phi\left(\frac{2}{\alpha_n}\right) \ .$$

One can readily check that

$$(5.70) \qquad \sum_{k=n+1}^{\infty} a_k \Phi\left(\frac{2}{\alpha_n}\right) \leq C(na_n \log n) \ \Phi\left(\frac{2}{\alpha_n}\right)$$

for some constant C, for $n \geq n_0$ for n_0 sufficiently large. Furthermore

$$(5.71) \qquad \sum_{k=1}^{n} a_n \Phi\left(\frac{2k}{\alpha_n n}\right) \leq 4 \sum_{k=1}^{n\alpha_n/2} a_k \frac{k^2}{n^2 \alpha_n^2} + na_{(n\alpha_n/2)} \Phi\left(\frac{2}{\alpha_n}\right)$$

$$\leq C_1 na_{(n\alpha_n/2)} \alpha_n + na_{(n\alpha_n/2)} \Phi\left(\frac{2}{\alpha_n}\right)$$

for some constant C_1 and $n \geq n_0$. Now since

$$a_{(n\alpha_n/2)} \sim a_n (\alpha_n)^{-1}$$

and $q > 1$, we see from (5.70) and (5.71) that

$$(5.72) \qquad E\Phi\left(\frac{|Y(u) - 1|}{\alpha_n}\right) < 1 \ , \quad \frac{1}{n-1} < u \leq \frac{1}{n}, \quad n \geq n_0 \ ,$$

if α in (5.68) is sufficiently large. By the definition of the Orlicz space norm $d_{X,\Phi}$, (5.72) implies (5.68). The reader may note that the argument of this paragraph remains valid for $p = 2$, $\beta \leq 0$. This fact will be used in the proof of Theorem 6.8.

One can write the metric entropy integral in (5.67) in terms of the non-decreasing rearrangement of $d_{X,\phi}(u,0)$ for $u \in [0,2\delta]$ which we will denote by $\overline{d_{X,\phi}(u,0)}$. This is outlined in Section 4 in (4.82) – (4.84). Thus (5.65) is equivalent to

$$(5.73) \qquad \int_0^{2\delta} \frac{\overline{d_{X,\phi}(u,0)}}{u(\log 1/u)^{1/p}} \, du < \infty$$

as long as $\delta < 1/4$. But we can take $\delta < n_0/2$ so that (5.68) is valid for all $u \in [0,2\delta]$. Since (5.68) gives a non-decreasing upper bound for $d_{X,\phi}(u,0)$ for $u \in [0,2\delta]$ it also gives an upper bound for $\overline{d_{X,\phi}(u,0)}$, $u \in [0,2\delta]$. Using this observation in (5.73) along with a change of variables we see that (5.73) and equivalently (5.58) hold if

$$(5.74) \qquad \sum_{n=n_0}^{\infty} \frac{\alpha_n}{n(\log n)^{1/p}} < \infty$$

for all n_0 sufficiently large. Thus we can see that (5.65) holds when $\varepsilon > 0$, i.e. for all those processes considered in Corollary 5.4 which are continuous.

We will now show that the integral in (5.65) is infinite for those processes considered in Corollary 5.4 which are unbounded a.s. By a change of variables (5.73) and therefore (5.65) hold if and only if

$$(5.75) \qquad \sum_{n=n_0}^{\infty} \frac{\overline{d_{X,\phi}(2^{-n},0)}}{n^{1/p}} < \infty \; .$$

We show that

$$(5.76) \qquad \overline{d_{X,\phi}(2^{-n},0)} \geq \beta_n \equiv \beta[n^{1/q}(\log n)^{1 + \varepsilon/p}]^{-1}$$

for some $\beta > 0$ independent of n for $n \geq n_0$ sufficiently large. This shows that (5.75) is false for $\varepsilon \leq 0$ in (5.61) which verifies the first sentence of this paragraph. Following the first part of this discussion it is enough to show that

$$d_{X,\Phi}(u,0) \geq \beta_n, \quad 2^{-n}\pi < u \leq 2^{-n+1}\pi$$

for $n \geq n_0$ for some n_0 sufficiently large. Clearly this will follow if we show that

(5.77)
$$E\Phi\left(\left|\frac{\sin ku}{\beta_n}\right|\right) \geq 1, \qquad 2^{-n}\pi < u \leq 2^{-n+1}\pi$$

for $n \geq n_0$ for some n_0 sufficiently large. To verify (5.77) we note that for u in the given range

(5.78)
$$E\Phi\left(\left|\frac{\sin ku}{\beta_n}\right|\right) \geq C_{p,\beta} \sum_{k=2^n}^{\infty} a_k \Phi\left(\left|\frac{\sin ku}{\beta_n}\right|\right)$$

$$\geq C_{p,\beta} \sum_{m=n}^{\infty} a_{2^{m+1}} \sum_{k=2^m}^{2^{m+1}-1} \Phi\left(\left|\frac{\sin ku}{\beta_n}\right|\right)$$

$$\geq C_{p,\beta} \sum_{m=n}^{\infty} 2^m a_{2^{m+1}} \Phi\left(\frac{1}{\beta_n}\left|\frac{1}{2^m}\sum_{k=2^m}^{2^{m+1}-1}\sin ku\right|\right) .$$

Now since $m \geq n \geq n_0$, if n_0 is sufficiently large

(5.79)
$$\frac{1}{2^m}\sum_{k=2^m}^{2^{m+1}-1}|\sin ku| \geq 1/4 .$$

Using (5.79) we see that the last line in (5.78)

(5.80)
$$\geq C_{p,\beta} \Phi\left(\frac{1}{4\beta_n}\right)\sum_{m=n}^{\infty} 2^m a_{2^{m+1}} \geq 1$$

if β in (5.76) is sufficiently small. Thus we have verified (5.77). This completes the demonstration of the assertion relating to (5.65).

We now repeat the above arguments as they pertain to (5.66). In this case $\Phi(u) = u^p$, $1 < p < 2$. Analagous to (5.68) we have that for $\frac{1}{n+1} < u \leq \frac{1}{n}$, and all $n \geq n_0$ sufficiently large

(5.81)
$$d_{X,p}(u,0) \leq \alpha\left[(\log n)^{1/q}(\log\log n)^{1+\beta/p+\epsilon/p}\right]^{-1} \equiv \alpha_n,$$

for some constant α sufficiently large. The entire argument in the paragraph containing (5.68) to (5.72) goes through verbatim for this value of Φ and α_n. Analagous to the relationship between (5.65) and (5.67) we also have that the integral in (5.66) is finite if and only if

$$(5.82) \qquad \int_0^\infty H_\Psi(\log N([-\delta,\delta],d_{X,p};\varepsilon))d\varepsilon < \infty \; .$$

The arguments of the paragraph containing (5.73) and (5.74) also show that (5.82) holds if

$$(5.83) \qquad \sum_{n=n_0}^\infty \frac{\alpha_n}{n^{\Psi^{-1}}(\log n)} < \infty$$

for α_n as given in (5.81). Thus we see that (5.66) holds for those processes in Corollary 5.4 which are continuous.

We now show that the integral in (5.66) is infinite for those processes in Corollary 5.4 which are unbounded a.s. Using the argument that led to (5.75) we now have that (5.66) holds if and only if

$$(5.84) \qquad \sum_{n=n_0}^\infty \frac{\overline{d_{X,p}(2^{-n},0)}}{\Psi^{-1}(n)} < \infty \; .$$

When $\Phi(u) = u^p$, we have, by exactly the same proof that led to (5.76) that

$$\overline{d_{X,p}(2^{-n},0)} \geq \beta_n \equiv \beta\left[n^{1/q}(\log n)^{1+\beta/p+\varepsilon/p}\right]^{-1},$$

for some $\beta > 0$ sufficiently small. Clearly for these values of β_n, (5.84) is infinite if $\varepsilon \leq 0$.

So far we have shown that the examples of Corollary 5.4 do not rule out the possibility that either (5.65) or (5.66) is a necessary and sufficient condition for the continuity a.s. of ξ-radial processes considered in Corollary 1.5. Let us now observe that $J(H_\Psi,d_{X,\Phi}) < \infty$, (see (1.56)) is not a necessary and sufficient condition for continuity

a.s. of the processes in Corollary 5.4 and hence for the more general

class of ξ-radial processes in Corollary 1.5. Consider (5.65) and

$J(H_\Psi, d_{X,\Phi})$ the metrics are the same but the functions of the metric

entropy are different. Thus these integrals are not equivalent for all

$\Psi(u) \sim u^p (\log u)^\beta$. (For example, instead of (5.74), $J(H_\Psi, d_{X,\Phi}) < \infty$ is

equivalent to

$$\sum_{n=m_0}^{\infty} \alpha_n (\log \log n)^{\beta/p} (n (\log n)^{1/p})^{-1} < \infty \; ,$$

for α_n as given in (5.68).)

Note that neither (5.65) nor (5.66) generalizes if $\Psi(u)$ is not

equivalent to a regularly varying function. In all of this our goal is

given Ψ to find a single integral condition for continuity that depends

only on the measure m, just as we did in the p-stable case in [MP2]. It

is possible that in these more general cases the function of the metric

entropy in the metric entropy integral also depends on m, if indeed there

is an integral condition for continuity, of the form of (1.7) with fixed

H and d, at all.

We now exhibit a different class of ξ-radial processes which leads

us to choose (5.66) rather than (5.65) in the conjecture in Chapter 1.

<u>Theorem 5.6</u>: Let $G = [0, 2\pi]$ so that $\Gamma = \{e^{ikt}, k \in N\}$. Let $\{X_N(t)\}_{t \in G}$

be a $\mathbb{C}^G$ valued stochastic process with characteristic functional given

by (1.5) in which

(5.85) $\Psi(u) \sim u^p (\log u)^\beta, \quad 1 < p < 2$

for u sufficiently large. Let the measure m_N in (1.5) satisfy

(5.86) $m_N \{e^{ikt}\} = N^{-1}$, $k = 1,\ldots,N$.

Then there exist constants c, C > 0 and independent of N, $\forall N \geq 3$, such

that

(5.87) $cH_\Psi (\log N) \leq E \sup_{t\varepsilon[0,2\pi]} |X_N(t)| \leq CH_\Psi (\log N)$

where H_Ψ is given in (1.9). (Note that in these cases, by regular

variation $H_\Psi(x) \sim x(\Psi^{-1}(x))^{-1}$ for x large.)

Proof: The lower bound for $E \sup_{t\varepsilon[0,2\pi]} |X(t)|$ is a direct consequence of

the proof of Theorem 1.1, part I.) given in Chapter 3. In this proof the

reader is refered to Theorem 2.9, [MP2]. The proof of this latter Theorem

along with the rest of the proof of Theorem 1.1, part I.) actually shows

that

(5.88) $E \sup_{t\varepsilon K} |X_N(t)| \geq$ Const. $J(H_\eta, d_{X,\Phi}, K)$

where $K \subset G$ is compact and where T, η and Φ are related to Ψ by

(1.10) and (1.11). The constant in (5.88) can depend on Ψ, η and Φ

but it is independent of N because in general it is independent of the

measure m.

    The relationship between Ψ, Φ and η is given in (1.10) and

(1.11). Let T be a convex function with $T(0) = 0$ such that $T \sim \Psi$.

We take $\eta(x) = T(x)$. If $\beta < 0$ in (5.85) we take $\Phi(x) = x^p$; if

$\beta \geq 0$ we take $\Phi(x) = T(x)$. The interesting point here is that for m_N

as given in (5.86) the results of this Theorem do not depend on the

particular function Φ that appears in the metric $d_{X,\Phi}$. Consider any

convex function $\Phi(x)$, $\Phi(0) = 0$, in (1.11). For the processes

$\{X_N(t)\}_{t\varepsilon[0,2\pi]}$ considered here, we have that

$$d_{X_N,\Phi}(u,0) = \inf \left\{ c > 0 : \frac{1}{N} \sum_{k=1}^{N} \Phi\left(\frac{|e^{iku}-1|}{c}\right) \le 1 \right\} .$$

By (5.78) and (5.79) with $a_k = N^{-1}$, $k = 1,\ldots,N$ and $C_{p,\beta} = 1$, we see that for $2^{-n}\pi < u < 2^{-n+1}\pi$ and $n \ge n_0$, for some n_0 sufficiently large,

$$(5.89) \qquad E \, \Phi\left(\frac{|\sin ku|}{\beta_n}\right) \ge \frac{1}{N} \sum_{m=n}^{[\log_2 N]} 2^m \, \Phi\left(\frac{1}{4\beta_n}\right)$$

and so for $n_0 \le n \le [\log_2 N] - 1$

$$(5.90) \qquad E \, \Phi\left(\frac{|\sin ku|}{\beta_n}\right) \ge \frac{1}{2} \, \Phi\left(\frac{1}{4\beta_n}\right).$$

Note that if $\frac{1}{2} \Phi\left(\frac{1}{4\beta_n}\right) \ge 1$ then $d_{X_N,\Phi}(u,0) \ge \beta_n$ for n in the indicated range. Thus we obtain that

$$(5.91) \qquad d_{X_N,\Phi}(u,0) \ge \frac{1}{4\Phi^{-1}(2)} \equiv \delta, \qquad \frac{2\pi}{N} \le u \le 2^{-n_0+1}\pi$$

where n_0 is independent of N and this holds for any convex function Φ. Let us now consider (5.88) with $k = [0, 2^{-n_0+1}\pi]$. It follows from (5.91) that whether $\Phi \sim \Psi$ or $\Phi(x) = |x|^p$,

$$(5.92) \qquad N(K, d_{X,\Phi}; \delta) \ge \left[\frac{N}{2\pi}\right] + 1 .$$

Therefore

$$J(H_\eta, d_{X_N,\Phi}, K) = \int_0^\infty H_\eta(\log N(K, d_{X_N,\Phi}; \varepsilon)) d\varepsilon \ge \delta H_\eta\left(\log \frac{N}{2\pi}\right)$$

and since δ is a constant we get

$$(5.93) \qquad E \sup_{t \in [0, 2\pi]} |X_N(t)| \ge \delta H_\eta\left(\log \frac{N}{2\pi}\right).$$

Since $\eta \sim \Psi$ we get the left hand side of (5.87).

To obtain the upper bound in (5.87) we note that if we had used the full strength of Lemma 4.2 in the proof of the Theorem 1.1 part II.) we would have obtained

$$(5.94) \qquad E \sup_{t \varepsilon [0,2\pi]} |X_N(t)| \le c[E|X_N(t_0)| + J(H_\eta, d_{X_N, \Phi}; [0, 2\pi])]$$

where C is a constant independent of N. One can see from the representations of ξ-radial processes given in (2.21) and (2.22) that $E|X_N(t_0)|$ is independent of N, and of course it is finite for the Levy transforms Ψ given in (5.85). Also, by essentially the same arguments that takes us from (4.82) to (4.84) we get

$$(5.95) \qquad J(H_\eta, d_{X_N, \Phi}, [0, 2\pi]) = \int_0^\infty H_\eta(\log N([0, 2\pi], \delta_{X_N, \Phi}, \varepsilon)d\varepsilon$$

$$\le \int_0^{2\pi} \frac{\overline{\sigma_N(s)}}{s\eta^{-1}(\log c/s)} \, ds = \int_{(2\pi)^{-1}}^\infty \frac{\overline{\sigma_N(1/u)}}{u\eta^{-1}(\log cu)} \, du$$

$$\le K \left[\int_{(2\pi)^{-1}}^N \frac{du}{u\eta^{-1}(\log cu)} + \int_N^\infty \frac{\overline{\sigma_N(1/u)}}{u\eta^{-1}(\log cu)} \, du \right]$$

where $\overline{\sigma_N(s)}$, the non-decreasing rearrangement of $d_{X_N, \Phi}(s, 0)$ is defined prior to (4.82). In what follows we will take $\eta(u) \sim \Psi(u)$ for u large. In this case, taking (5.85) into account we see that the first integral in the last line of (5.95)

$$(5.96) \qquad \le C' \frac{\log N}{\Psi^{-1}(\log N)} \quad ,$$

where C' is a constant independent of N. In order to evaluate the second integral in the last line of (5.95) we note that for $u \le \frac{1}{N}$ and any convex function $\Phi(0) = 0$ we have

$$E\Phi\left(\left|\frac{Y(u)-1}{C(u)}\right|\right) \leq \frac{1}{N}\sum_{k=1}^{N}\Phi\left(\frac{2k}{C(u)u}\right) \leq \Phi\left(\frac{2N}{C(u)u}\right)$$

Since there exists an $a < \infty$, independent of N, $C(u) = \frac{aN}{u}$ implies $\Phi\left(\frac{2N}{C(u)u}\right) \leq 1$ we see that

(5.97) $$d_{X_N,\Phi}(u,0) \leq \frac{aN}{u}, \qquad u > \frac{1}{N}.$$

Using (5.97) we see that the second integral in the last line of (5.95)

(5.98) $$\leq Ka\int_N^\infty \frac{N\,du}{u^2\Psi^{-1}(\log cu)} \leq \frac{C'}{\Psi^{-1}(\log N)}$$

where C' is a constant independent of N. Finally by (5.94), (5.96) and (5.97) we get

$$E\sup_{t\in[0,2\pi]}|X_N(t)| \leq C\left[1 + \frac{\log N}{\Psi^{-1}(\log N)}\right]$$

which gives us the right side of (5.87). However there remains one point to justify, that is, that we can take $\Psi(u) \sim \eta(u)$ for u large. But this is easy to see. If $\beta \geq 0$ in (5.85) we can take all three functions, $T = \eta = \Phi$ to be some convex extension of $u^p(\log u)^\beta$; if $\beta \leq 0$ in (5.85) we can take $T = \eta$ to be convex extentsions of $u^p(\log u)^\beta$ and take $\Phi(x) = x^p$. This completes the proof of Theorem 5.6.

Remark 5.7: The range $1 < p < 2$ in (5.85) is chosen for convenience. The same proof we used actually shows that the lower bound in (5.87) is valid for

(5.99) $$\Psi(u) \sim u^p(\log)^\beta, \qquad 0 < p \leq 2,$$

although $H_\Psi(x)$ is not comparable to $x(\Psi^{-1}(x))^{-1}$ when $p = 1$. Furthermore $H_\Psi(x) < \infty$ when $p < 1$ and, since Ψ is a Levy transform,

$H_\psi(x) = o(x^{1/2})$, as we have remarked above. The upper bound in (5.87)

also remains valid when $p = 1$ and $p = 2$ in (5.99). This can be

obtained from Theorem 6.6 when $p = 2$. (Following the paragraph

containing 5.94-5.98.) When $p = 1$ we have no upper bound in terms of

the metric entropy but in the special cases considered in (5.99) we can

use the classical techniques employed in the paragraph containing

(5.40)-(5.46) to show that

$$(5.100) \qquad E \sup_{t \varepsilon [0,2\pi]} |X_N(t)| \leq c [NE|\xi| I_{[|\xi|>b_N]}$$

$$+ (NE|\xi|^2 I_{[|\xi|<b_N]})^{1/2} (\log N)^{1/2}]$$

where $E \exp i\lambda\xi = \exp - N^{-1}\psi(\lambda)$ and C is a constant independent of

N. (We use (5.43) with $\{\xi_k'\}$ replaced by $\{\xi_k\}$ and (5.44) which is the

classical upper bound for subgaussian series; see e.g. Theorem 1.1,

Chapter I and Lemma 1.1, Chapter VII, [MP1].) Taking $b_N = (\psi^{-1}(\log N))^{-1}$

in (5.100) one obtains the right side of (5.87) whenever $1 \leq p < 2$.

As we have just shown the upper bound in (5.87) can be obtained by

classical methods, (at least for $1 \leq p < 2$). This is not the case for

the lower bound. The classical approach to the lower bound would be to

use Theorem 5.1 but the right side of (5.2) is not even unbounded as

$N \to \infty$. Gilles Pisier has a clever proof (unpublished) of the fact that

$$(5.101) \qquad E \sup_{t \varepsilon [0,2\pi]} |X_N(t)| \geq C [\sum_{1}^{\log N} F^{-1}(k)]$$

for F^{-1} as defined in Lemma 2.1 in terms of the Levy measure of ψ,

where C is a constant independent of N. When $1 \leq p < 2$

$F^{-1}(k) \sim (\psi^{-1}(k))^{-1}$ so (5.88) gives the lower bound in (5.87) in these

cases. However the right side of (5.101) is too small when $p = 2$. As

far as we know the lower bound in (5.87), which, as we remarked is valid for $0 < p \leq 2$, actually requires Theorem 1.1, part I.)

Let us note that the method of proof of Theorem 5.6 also shows that for $\Psi(u)$ as given in (5.85)

$$(5.102) \quad c(\log N)^{1/q} \leq \int_0^\infty (\log N[0,2\pi],d_{X_N,\Phi};\varepsilon)^{1/q}d\varepsilon \leq C(\log N)^{1/q}$$

for constants c and C independent of N, regardless of the value of β, (here $\frac{1}{p} + \frac{1}{q} = 1$). However since

$$\frac{\log N}{\Psi^{-1}(\log N)} \sim (\log N)^{1/q}(\log \log N)^{\beta/p}$$

we see from Theorem 5.6 and particularly from (5.93) that the integral in (5.102), which is (5.65), does not give the correct rate of growth, as a function of N, for the class of processes $\{X_N(t)\}_{t \in G}$ considered in Theorem 5.6.

Remark 5.8: In (5.87) c and C, although independent of N, are not independent of Ψ. Thus (5.87) is not homogeneous in the sense that if we replace $X_N(t)$ by $aX_N(t)$ the upper and lower bounds in (5.87) are not simply multiplied by a. For clarification note that in the definition of ξ-radial processes the measure m is always taken to be a probability measure. Therefore to consider aX_N as a ξ-radial process we must replace $\Psi(x)$ in (5.85) by $\Psi(ax)$ and, if we denote $\Psi(x)$ by Ψ_1 and $\Psi(ax)$ by Ψ_2, obviously $H_{\Psi_1}(\log N)$ is not equivalent to $aH_{\Psi_2}(\log N)$.

Since (5.87) is not homogeneous we can not use standard techniques to show that (5.65) is not a necessary and sufficient condition for continuity of ξ-radial radial processes. Indeed we do not have an example of a ξ-radial process, for which $\Psi(u)$ is regularly varying with index $1 < p \leq 2$, that shows this. However when $\Psi(u)$ is regularly varying of index 1 we can exhibit a continuous ξ-radial process for which

the appropriate extension of the integral in (5.65) is infinite. This
example which will be given in Chapter 6 and the fact that (5.66) reflects
the correct rate of growth of $X_N(t)$ in Theorem 5.6 is why we chose
(5.66) rather than (5.65) as (1.74) in the conjecture in Chapter 1. This
appears to be a delicate point because both the integral conditions (5.65)
and (5.66) work for the processes considered in Corollary 5.4.

Obtaining upper and lower bounds for the expected value of the
suprema of ξ-radial processes that show their dependence on Ψ should be
an important next step in the study of ξ-radial processes.

We now turn our attention back to random Fourier series. We shall
show that there is no integral condition for continuity for the processes
considered in Corollary 1.6. In this Corollary we considered two
integrals

(5.103) $$\int_0^\infty (\log N([0,2\pi], d_{Y,\phi};\varepsilon)^{1/q} d\varepsilon$$

and

(5.104) $$\int_0^\infty H(\log N([0,2\pi], d_{Y,\phi};\varepsilon) d\varepsilon$$

where H is given in (1.62). We shall first show that (5.103) is finite
for the continuous processes considered in Corollary 5.5 and infinite for
those which are discontinuous. Next we consider a simple class of
lacunary random Fourier series and show that, this time, (5.104) is finite
for the continuous ones and infinite for the ones which are
discontinuous. Since these integrals are not equivalent when $\beta \neq 0$ we
see that Corollary 1.6 is actually "best possible" and that there is no
integral condition for continuity of the random Fourier series considered
in Corollary 1.6 except when $\beta = 0$.

Consider the examples in Corollary 5.5 and the integral in (5.103).
We follow exactly the same argument used to relate (5.65) with the

examples in Corollary 5.4 except that we consider the metric $d_{Y,\phi}$
instead of $d_{X,\phi}$. Analogous to (5.68) and (5.76) we show that for
$\frac{1}{n+1} < u \leq n$ for all $n \geq n_0$ sufficiently large

$$(5.105) \qquad d_{Y,\phi}(u,0) \leq \tilde{\alpha}[(\log n)^{1/q} (\log \log n)^{1+\epsilon}]^{-1} \equiv \tilde{\alpha}_n$$

and for $2^{-n}\pi < u \leq 2^{n+1}\pi$ and all $n \geq n_0$ sufficiently large

$$(5.106) \qquad d_{Y,\phi}(u,0) \geq \tilde{\beta}(n^{1/q}(\log n)^{1+\epsilon})^{-1} \equiv \tilde{\beta}_n \ .$$

The rest of the argument proceeds exactly as the one for the ξ-radial
processes.

To obtain (5.105) it is enough to show that for $\frac{1}{n+1} < u \leq \frac{1}{n}$, $n \geq n_0$

$$(5.107) \qquad \sum_{k=1}^{\infty} \phi\left(\frac{2|a_k \sin ku|}{\tilde{\alpha}_n}\right) \leq 1 \ .$$

We have

$$(5.108) \qquad \sum_{k=1}^{\infty} \phi\left(\frac{2|a_k \sin ku|}{\tilde{\alpha}_n}\right) \leq \sum_{k=1}^{n} \phi\left(\frac{2|a_k|k}{n\tilde{\alpha}_n}\right)$$

$$+ \sum_{k=n+1}^{\infty} \phi\left(\frac{2|a_k|}{\tilde{\alpha}_n}\right) \ .$$

For n sufficiently large the arguments of the functions ϕ are very
small so we can take $\phi(\lambda) = \lambda^p(\log 1/\lambda)^{\beta}$, $\lambda < 1/\epsilon$. We have, for u
sufficiently large

$$(5.109) \qquad \sum_{k=n+1}^{\infty} \phi\left(\frac{2|a_k|}{\tilde{\alpha}_n}\right) \leq C \sum_{k=n+1}^{\infty} \left|\frac{a_k}{\tilde{\alpha}_n}\right|^p \left(\log \frac{\tilde{\alpha}_n}{a_k}\right)^{\beta}$$

$$\leq \frac{C_1}{|\tilde{\alpha}_n|^p} \sum_{k=n+1}^{\infty} |a_k|^p(\log k)^{\beta} \leq 1/2$$

if $\tilde{\alpha}$ in (5.105) is taken to be sufficiently large. Also

(5.110) $\sum\limits_{k=1}^{n} \phi\left(\dfrac{2|a_k|k}{n\tilde{\alpha}_n}\right) \leq C\, n\phi\left(\dfrac{2a_n}{\tilde{\alpha}_n}\right) \sim Cn\left|\dfrac{a_n}{\tilde{\alpha}_n}\right|^p (\log n)^\beta \leq 1/2$

for $\tilde{\alpha}$ sufficiently large. Combining (5.109) and (5.110) we get (5.107) and hence (5.105).

To obtain (5.106) we note that

(5.111) $\sum\limits_{n=0}^{\infty} \phi\left(\dfrac{2|a_k \sin ku|}{\tilde{\beta}_n}\right) \geq \sum\limits_{k=2^n}^{\infty} \phi\left(\dfrac{2|a_k \sin ku|}{\tilde{\beta}_n}\right)$

$$= \sum\limits_{m=n}^{\infty} \sum\limits_{k=2^m}^{2^{m+1}-1} \phi\left(2a_{2^{m+1}}\left|\dfrac{\sin ku}{\tilde{\beta}_n}\right|\right)$$

$$\geq \sum\limits_{m=n}^{\infty} 2^m\, \phi\left(\dfrac{2a_{2^{m+1}}}{\tilde{\beta}_n}\left(\dfrac{1}{2^m}\sum\limits_{k=2^m}^{2^{m+1}-1}\sin ku\right)\right)$$

$$\geq \sum\limits_{m=n}^{\infty} 2^m \phi\left(\dfrac{a_{2^{m+1}}}{2\tilde{\beta}_n}\right)$$

$$\geq d_{p,\beta}\left|\dfrac{1}{\tilde{\beta}_n}\right|^p \sum\limits_{m=n}^{\infty} 2^m |a_{2^{m+1}}|^p\, m^\beta \geq 1$$

if $\tilde{\beta}$ in (5.106) is sufficiently small. (Here $d_{p,\beta}$ is a constant depending on p and β and, as above, we note that the arguments of ϕ can be made as small as we wish by taking n_0 sufficiently large. Thus we can let $\phi(\lambda) = \lambda^p(\log 1/\lambda)$, $\lambda < 1/e$ as above. We also use (5.79).) Note that (5.111) implies (5.106)

Now that we verified (5.105) and (5.106) we complete the argument by noting that (5.103) is the same as (5.73) with $d_{Y,\Phi}$ replaced by $d_{Y,\phi}$. Thus (5.103) is finite if (5.74) is finite (with $\tilde{\alpha}_n$ replacing α_n) and infinite if (5.75) is infinite (with $\tilde{\beta}_n$ replacing β_n (see (5.76)). In other words (5.103) is finite when $\varepsilon > 0$ and infinite when $\varepsilon \leq 0$ and this corresponds to the continuous and unbounded processes in Corollary 5.5.

Now with ϕ, ψ and H still as in Corollary 1.6 let us consider the lacunary random Fourier series

$$(5.112) \qquad Y(t) = \sum_{k=0}^{\infty} b_k \xi_k e^{i2^k t}, \quad t \in [0, 2\pi]$$

where, as usual, $\{\xi_k\}_{k=1}^{\infty}$ are i.i.d. copies of ξ and ξ and ψ are related as in (1.17). We consider the class of examples corresponding to

$$(5.113) \qquad b_k = (k(\log k)^a)^{-1}, \quad k \geq 10$$

$$b_k = 0 \qquad k = 1, \ldots, 9$$

where $-\infty < a < \infty$. Since $\{2^k\}_{k=0}^{\infty}$ is a lacunary set we have by Corollary 1.6, II.) for those ϕ and ψ for which $\beta < 0$ that if the integral in (5.104) is finite then $a > 1$. We will now show that if $a > 1$ in (5.113) then the integral in (5.104) is finite for all β.

Following the argument in the paragraph containing (4.82)-(4.86) we see that (5.104) is finite if and only if

$$(5.114) \qquad \sum_{n=n_0}^{\infty} \frac{\overline{d_{Y,\phi}(2^{-n}, 0)}}{n^{1/p} (\log n)^{\beta/p}} < \infty .$$

We will show that for $2^{-n-1} < u \leq 2^{-n}$ and $n \geq n_0$ sufficiently large

$$(5.115) \qquad d_{Y,\phi}(u, 0) \leq \tilde{\alpha}(n^{1/q}(\log n)^{a-\beta/p})^{-1} \equiv \tilde{\alpha}_n$$

for some $\tilde{\alpha}$ sufficiently large. Thus for $a > 1$, (5.114) and hence (5.104) are finite.

To obtain (5.115) it is enough to show that for $2^{-n-1} < u \leq 2^{-n}$, $n \geq n_0$

(5.116)
$$\sum_{k=0}^{\infty} \phi\left(\frac{2|a_k \sin 2^k u|}{\tilde{\alpha}_n}\right) \leq 1 \ .$$

We have

$$\sum_{k=0}^{\infty} \phi\left(\frac{2|a_k \sin 2^k u|}{\tilde{\alpha}_n}\right) \leq \sum_{k=1}^{n} \phi\left(\frac{2|a_k|}{\tilde{\alpha}_n 2^{n-k}}\right) + \sum_{k=n+1}^{\infty} \phi\left(\frac{2|a_k|}{\tilde{\alpha}_n}\right) \ .$$

For $k \geq n \geq n_0$, sufficiently large, $\log(\tilde{\alpha}_n/a_n) \sim \log k$, thus we have

(5.117)
$$\sum_{k=n+1}^{\infty} \phi\left(\frac{2|a_k|}{\tilde{\alpha}_n}\right) \leq c \sum_{k=n+1}^{\infty} \left(\frac{a_n}{\tilde{\alpha}_n}\right)^p \left(\log \frac{\tilde{\alpha}_n}{a_k}\right)^\beta$$

$$\leq \frac{C_1}{|\tilde{\alpha}_n|^p} \sum_{k=n+1}^{\infty} |a_k|^p (\log k)^\beta \leq 1/2$$

if $\tilde{\alpha}$ in (5.115) is taken to be sufficiently large. Also by monotonicty, for n sufficiently large,

(5.118)
$$\sum_{k=1}^{n} \phi\left(\frac{2|a_k|}{\tilde{\alpha}_n 2^{n-k}}\right) \leq n\phi\left(\frac{2a_n}{\tilde{\alpha}_n}\right)$$

$$\leq C'n \left|\frac{a_n}{\tilde{\alpha}_n}\right|^p (\log n)^\beta \leq 1/2$$

for $\tilde{\alpha}$ sufficiently large. Combining (5.117) and (5.118) we get (5.116) and thus (5.115). This shows that (5.104) is finite if and only if the processes considered in (5.112) and (5.113) are continuous. Thus this class of examples and those of Corollary 5.5 establish our assertion about the absence of an integral condition for the processes considered in Corollary 1.6 when $\beta \neq 0$.

Remark 5.9: Lacunary random Fourier series which are ξ-radial processes have also been studied. One can use Theorem 1.7 to generate examples by choosing the measure to assign mass to $\{e^{i2^k t}\}_{k=0}^{\infty}$. We have studied such examples when the measure is sufficiently smooth and found that both (5.65) and (5.66) hold for those processes which are continuous and neither (5.65) nor (5.66) hold for those processes which are unbounded. (In these examples Ψ and Φ are as given in Corollary 1.5). We are not including the details because the arguments are exactly the same as those used in studying the examples of Corollary 5.4).

6. PROCESSES FOR WHICH THE LEVY TRANSFORMS OR THE LOGARITHMS OF THE
CHARACTERISTIC FUNCTIONS ARE REGULARLY VARYING WITH INDEX 1 OR 2.

The processes considered in this Chapter are intriguing because, in some sense, thay are at the boundary of the interesting infinitely divisible processes. Thus when $\Psi(x) = \psi(x) = x^2$ one has Gaussian processes and at the other extreme, that is when $p = 1$, there are classes of these processes for which the continuity conditions are trivial. The results on necessary conditions for continuity obtained in Theorem 1.1 and 1.2 apply to these processes but the results on sufficient conditions for continuity in these Theorems do not. This is because the Tauberian Theorem of [BT] which was used repeatedly in Chapter 5 is no longer valid in the case $p = 2$ and in the case $p = 1$ the techniques we used to prove continuity are not very precise. Obtaining necessary and sufficient conditions for continuity, even for 1-stable random Fourier series, remains an open problem.

Our consideration of the case $p = 1$ consists mainly of obtaining examples. These examples support our choice of (5.66) over (5.65) in the conjecture in Chapter 1. Sufficient conditions for continuity in the case $p = 2$ are given in Theorems 6.5 and 6.7 for ξ-radial processes and in Theorem 6.9 for random Fourier series. Examples are also obtained in the case $p = 2$. Our approach is the same as in Chapter 5 but the computations are more delicate.

It is convenient to study the two cases of indices of regular variation separately. We begin with those of index 1 and examine first ξ-radial processes.

Theorem 6.1: Let $G = [0, 2\pi]$ so that $\Gamma = \{e^{ikt}, k \in \mathbb{N}\}$. Let $\{X(t)\}_{t \in G}$ be a $\mathbb{C}^G$ valued stochastic process with characteristic functional given by (1.5) with $\Psi(u) \sim u(\log u)^\beta$, $-1 < \beta$, for u sufficiently large. Let the measure m in (1.5) satisfy

(6.1) $m\{e^{ikt}\} = C_\beta[k(\log k)(\log \log k)^{2+\beta}(\log \log \log k)^{1+\varepsilon}]^{-1} \equiv a_k$

for $k \geq 20$, $m\{e^{ikt}\} = C_\beta$, $0 \leq k \leq 20$, where C_β is such that $|m| = 1$. Then if $\varepsilon > 0$ in (6.1), $\{X(t)\}_{t \in G}$ has a version with continuous sample paths. However, if $\varepsilon \leq 0$, $\sup_{t \in G}|X(t)|$ is unbounded a.s.

Proof: We first obtain the condition for unboundedness. Following the proof of Theorem 1.7 and in particular (5.33) we see that $\sup_{t \in G}|X(t)|$ is unbounded a.s. if

(6.2) $$\sum_{i=1}^{\infty} m_i \left(\sum_{j=j_0}^{[\frac{1}{m_i}]+1} \frac{1}{\Psi^{-1}(j)} \right) = \infty$$

for all $j_0 > 0$ where $m_i = m\{e^{ikt}: 2^i < k \leq 2^{i+1}\}$.
By (6.1)

(6.3) $m_i \sim \left(i(\log i)^{2+\beta}(\log \log i)^{1+\varepsilon}\right)^{-1}$, $i \geq 10$,

and since $\Psi^{-1}(j) \sim \dfrac{j}{(\log j)^\beta}$, $\beta > -1$ we see that (6.2) holds as long as $\varepsilon \leq 0$. (Note that the computation is not exactly the same as that of the proof of Theorem 1.7 because $\Psi^{-1}(j)$ is regularly varying with index 1.)

We now prove the assertion pertaining to continuity. Following the proof of Theorem 1.7 it is enough to show that the series

(6.4) $H(t) = \sum_{k=0}^{\infty} \xi_k e^{ikt}$, $t \in [0, 2\pi]$

has continuous sample paths when $\{\xi_k\}$ is independent and satisfies

$$E \, e^{i\lambda \xi_k} = \exp - a_k \Psi(|\lambda|) \ .$$

Also, let

$$b_k = (\log k \, (\log \log k)^2)^{-1} \ , \ k \geq k_0 > 0$$

for some k_0 sufficiently large. Still following the proof of Theorem 1.7 we define

$$\xi_k' = \xi_k \, I_{[\, |\xi_k| \, \leq \, 1]} \ , \qquad \forall k \geq 1.$$

In order to show that the series in (6.4) converges uniformly a.s. it is enough to show that

(6.5)
$$\sum_{k=1}^{\infty} P(|\xi_k| > 1) < \infty \ ,$$

(6.6)
$$\sum_{k=k_0}^{\infty} E |\xi_k'| \, I_{[\, |\xi_k'| \, > \, b_k]} < \infty$$

and

(6.7)
$$\sum_{n=k_0}^{\infty} \frac{\left(\sum_{k=n}^{\infty} E |\xi_k|^2 \, I_{[\xi_k \leq b_k]} \right)^{1/2}}{n(\log n)^{1/2}} < \infty \ .$$

By (5.45) and the fact that $\Psi(|\lambda|)$ is regularly varying at infinitely we see that

(6.8)
$$P(|\xi_k| > u) \leq 7 a_k u \int_0^{1/u} \Psi(|\lambda|) \, d\lambda$$

$$\leq c a_k \, \Psi(|\tfrac{1}{u}|)$$

for $u \geq 1$ and some constant c. This inequality shows that (6.5) is satisfied for $\varepsilon > 0$. Also, using (6.8) we see that

(6.9)
$$E |\xi_k'| \, I_{[\, |\xi_k'| > b_k]} \leq c a_k \int_1^{1/b_k} \frac{\Psi(|s|)}{s^2} \, ds$$

$$= c a_k \int_1^{1/b_k} \frac{(\log s)^\beta}{s} \, ds \sim c a_k (\log 1/b_k)^{\beta+1} \ .$$

(Here again the fact that $\Psi(|s|)$ is regularly varying of index 1 adds a factor of log $1/b_k$ to the calculation.) It follows from (6.9) that (6.6) is satisfied for $\varepsilon > 0$. Finally we note that by (6.8) we also have that

$$E|\xi_k|^2 \; I_{[\xi_k \leq b_k]} \leq ca_k \int_{1/b_k}^{\infty} \frac{\Psi(s)}{s^3} \; ds$$

$$\leq c'a_k b_k^2 \; \Psi\left(|\frac{1}{b_k}|\right)$$

$$\sim c'a_k b_k \left(\log \frac{1}{b_k}\right)^{\beta} .$$

Therefore

$$\sum_{k=n}^{\infty} E|\xi_k|^2 \; I_{[\xi_k \leq b_k]} \sim \left[(\log n)(\log \log n)^4 (\log \log \log n)^{1+\varepsilon}\right]^{-1}$$

and this shows that (6.7) is satisfied for all ε. This completes the proof of Theorem 6.1.

We will now show that (1.74), which in this case is the same as (5.66), holds for those processes in Theorem 6.1 which are continuous but not for those which are unbounded. On the other hand we will exhibit discontinuous processes for which

$$(6.10) \qquad \int_0^{\infty} (\log \log N([0, 2\pi], d_{X,\Phi}; \; \varepsilon) d\varepsilon < \infty .$$

This integral is the proper version of (5.65) when $\Phi(x)$ is regularly varying at infinity with index 1. Thus the examples in Theorem 6.1 support our choice of (5.66) over (5.65) in the conjecture in Chapter 1.

We first verify the assertion relating to the integral in (6.10). Parallel to (5.68) we first show that for $\frac{1}{n+1} < u \leq \frac{1}{n}$ and all $n \geq n_0$ sufficiently large

$$(6.11) \qquad d_{X,\Phi}(u, \; 0) \leq c \; \frac{(\log \log \log n)^{\beta - 1 - \varepsilon}}{(\log \log n)^{1+\beta}} \equiv \alpha_n$$

where $c > 0$ is some constant independent of $n \geq n_0$ for n_0 sufficiently
large. To obtain (6.11) we use the calculations given between (5.69) and
(5.71). One can check that for n sufficiently large

$$(6.12) \qquad \sum_{k=1}^{n} a_k \, \Phi\left(\frac{2k}{\alpha_n}\right) = o\left(\sum_{k=n+1}^{\infty} a_k \, \Phi\left(\frac{2}{\alpha_n}\right)\right)$$

and, with $c > 0$ large enough,

$$(6.13) \qquad \sum_{k=n+1}^{\infty} a_k \, \Phi\left(\frac{2}{\alpha_n}\right) < 1 \; .$$

Following the argument given in the paragraph containing (5.73) and
(5.74), we see that (6.10) holds if

$$\sum_{n=n_0}^{\infty} \frac{\alpha_n}{n(\log n)} < \infty$$

which is the case whenever $\beta > 0$, irregardless of the value of ε.

On the other hand for $\frac{1}{n+1} < u \leq \frac{1}{n}$

$$d_{X,1}(u,0) \sim \sum_{k=n+1}^{\infty} a_k \sim \left((\log \log n)^{1+\beta}(\log \log \log n)^{1+\varepsilon}\right)^{-1} \equiv \beta_n \; .$$

The integral condition in (1.74) or equivalently (5.66) is

$$(6.14) \qquad \int_{0}^{\infty} \left(\log \log N \left([0,2\pi],d_{X,1};\varepsilon\right)\right)^{1+\beta} d\varepsilon < \infty \; .$$

Following the argument given in the paragraph containing (5.73) and (5.74)
we see that (6.14) holds if and only if

$$(6.15) \qquad \sum_{n} \frac{\alpha_n}{n(\log n)(\log \log n)^{-\beta}} < \infty$$

which, of course, it is whenever $\varepsilon > 0$.

The next Theorem gives us examples of random Fouries series.

Theorem 6.2: Let $G = [0, 2\pi]$ so that $\Gamma = \{e^{ikt}, k \in \mathbb{N}\}$. Let

$\psi(\lambda) \sim \lambda(\log 1/\lambda)^{\beta}$, for $\lambda > 0$ sufficently close to zero; let ξ be as given

in (1.17) and consider

$$(6.16) \qquad\qquad Y(t) = \sum_{k=0}^{\infty} a_k \xi_k \, e^{ikt}, \quad t \in G$$

where $\{\xi_k\}_{k=0}^{\infty}$ are i.i.d. copies of ξ. Let

$$(6.17) \qquad\qquad a_k = \left[k(\log k)^{1+\beta} (\log \log k)^{2+\epsilon} \right]^{-1}$$

$k \geq 10$, otherwise let $a_k = 1$. Then if $\epsilon > 0$ in (6.17) the series in (6.16)

converges uniformly a.s. whereas if $\epsilon \leq 0$, $\sup\limits_{t \in G} |Y(t)|$ is unbounded a.s.

Proof:. Since this proof parallels the proof of Theorem 1.8 we will give

fewer details. Because $\psi(u)$ is regularly varying at zero we have that

$$(6.18) \qquad P(|\xi| > u) \sim C \, \psi\left(\frac{1}{u}\right) \sim C \, \frac{(\log u)^{\beta}}{u}, \quad u \geq u_0$$

for some constant C and u_0 sufficiently large. Following (5.3) we let

$$(6.19) \qquad\qquad \alpha_i = \inf\left\{ \lambda > 0: \sum_{k=2^i}^{2^{i+1}-1} P(|a_k \xi_k| > \lambda) \leq 1 \right\}.$$

We see that

$$\alpha_i \sim \left[i(\log i)^{2+\epsilon} \right]^{-1}$$

for i sufficiently large. For these values of $\{\alpha_i\}$ the sum in (5.4)

converges iff $\epsilon > 0$. Thus we get the result on unboundedness of the

series in (6.16). (Note that in this case both series on the right in

(5.2) converge!)

    It follows from (6.18) that $\sum\limits_{k=0}^{\infty} P(|a_k \xi_k| > 1) < \infty$. Therefore, as in

the proof of Theorem 1.8, we only need to show (5.56) and (5.57) with

$\{\xi_k\}$ replaced by $\{\xi_k I_{[\xi_k \leq 1/a_k]}\}$. The choice of $\{b_k\}$ is open to us.

We take $b_k = \dfrac{c_k}{a_k}$ where $c_k = (\log k)^{-1}$. One can now check that with

$\{a_k\}$ as given in (6.11) and with these values of $\{b_k\}$, (5.56) and (5.57)

are satisfied. This completes the proof of Theorem 6.2.

In direct analogy with the relationship between (5.103) and the

examples in Corollary 5.5 one can check that

$$(6.20) \qquad \int_0^\infty \Big(\log \log N([0,\, 2\pi]),\, d_{Y,\phi};\, \varepsilon\Big)\ d\varepsilon < \infty$$

for processes in Theorem 6.2 which are continuous but not for the unboun-

ded processes. We take $\phi(x) \sim \psi(x)$ for x near zero in the examples and

define $d_{Y,\phi}$ as in (1.21) even for $\beta < 0$. Note that (6.20) is the

proper version of (5.103) when $p = 1$. However, still in analogy with the

random Fourier series considered in Chapter 5, (6.20) can be infinite for

continuous lacunary random Fourier series. This is easy to check.

Theorem 1.1 contians necessary conditions for continuity of ξ-radial

processes when the Levy transform is regularly varying at infinity with

index 1. In fact when $\Psi(x) \sim T(x)$ and $T(x)$ satisfies (1.32) with

$p = 1$ and (1.33) the integral in (1.74) itself is a necessary condition

for continuity. To clarify this we will present an analogue of the

necessary part of Corollary 1.5 when Ψ is regularly varying with index 1.

<u>Theorem 6.3.</u> Let $\{X(t)\}_{t\epsilon K}$ be a $\mathbb{C}^K$ valued stochastic process with

characteristic functional given by (1.3) or (1.5) with $\Psi(u) \sim u(\log u)^\beta$,

for u sufficiently large. Let $\Phi: R^+ \rightarrow R^+$ be non-decreasing and satisfy

$$(6.21) \qquad\qquad \Phi(u) \sim u\,(\log u)^\beta\,, \qquad u \geq u_0 \geq 10$$

for some u_0 sufficiently large. For s,t ϵ K let $d_{X,\Phi}(s,t)$ be given as

in (1.6).

I.) Let $\beta \geq 0$. Then

(6.22) $$\int_0^\infty (\log \log N(K, d_{X,1}; \varepsilon))^{1+\beta} d\varepsilon = \infty$$

implies $\sup_{t \in K} |X(t)|$ is unbounded a.s.

II.) Let $-1 < \beta < 0$. Then

(6.23) $$\int_0^\infty (\log \log N(K, d_{X,\Phi}; \varepsilon))^{1+\beta} d\varepsilon = \infty$$

implies $\sup_{t \in K} |X(t)|$ is unbounded a.s.

Proof: I.) follows from Theorem 1.1 with $\Psi \sim T \sim \eta$ at infinity and with $\Phi(x) = x$ in (1.11). Note that in this case $T(x)$ satisfies (1.33) with $p = 1$. II.) follows from Part I.) of Theorem 1.1, which remains valid as stated if the statement Φ convex (see the paragraph containing (1.6)) is replaced by, "Φ non-decreasing and there exists a $k < \infty$ such that

(6.24) $$\Phi\left(\frac{c}{k}\right) \le \frac{1}{2} \Phi(c) \qquad \forall c \ge 0."$$

($d_{X,\Phi}$ is still defined as in (1.6) although it is no longer necessarily a pseudo-norm.) The reason for this is that the proof of Theorem 1.1 part I.) does not use the fact that $d_{X,\Phi}$ is a metric. The function $\tilde{\Phi}(x)$ defined in (3.6) does not have to be convex. In fact one can simply take $\tilde{\Phi}$ to satisfy (3.6) and to be equal to $\Phi(x)$ for $x \ge x_0'$. The statement in (3.15) can be obtained using (6.24) and, lastly, the critical "Proof of Lemma 2.1" in [MP2] uses the fact that d_ω is a metric but not that $d_{X,\Phi}$ is one. Note that in using Theorem 1.1 part I.) in this case we take $\Psi \sim T \sim \eta \sim \Phi$.

Let us briefly mention the problem of finding sufficient conditions for continuity of ξ-radial processes for which the Levy transform $\Psi(u)$ is regularly varying at infinity with index 1. The techniques employed in this paper are not very precise. This was already evident in [MP2] because even for 1 stable processes we could not obtain what we believed was the right condition for continuity. The critical step, based on Lemma 2.3 of [MP1], which is used repeatedly in this paper, seems to yield an extra power of the logarithm in the 1-stable case. It does not seem worthwhile to discuss the more general situation (i.e. $\Psi(u)$ regularly varying at infinity with index 1) until we can clear up the 1-stable case.

Our next result is an anlogue of the necessary part of Corollary 1.6 when ψ is regularly varying at zero with index 1.

<u>Theorem 6.4</u>: Let $\psi(\lambda) \sim \lambda(\log 1/\lambda)^\beta$ for $\lambda > 0$, sufficiently close to zero. Let ξ be given as in (1.17) and consider the random Fourier series $\{Y(t)\}_{t \in K}$ as defined in (1.18) where $\{\xi_\gamma\}_{\gamma \in A}$ are i.i.d. copies of ξ. Let $\phi : \mathbb{R}^+ \to \mathbb{R}^+$ be a non-decreasing function satisfying

(6.25) $$\phi(\lambda) \sim \lambda\big(\log(e + 1/\lambda)\big)^\beta, \ \lambda \leq \lambda_0$$

for some $\lambda_0 > 0$. For $s,t \in K$ define $d_{Y,\phi}(s,t)$ as in (1.21) for this function ϕ.

I.) Let $\beta \leq 0$. Then

(6.26) $$\int_0^\infty \big(\log \log N(K, d_{Y,\phi}(s,t); \varepsilon)\big) \, d\varepsilon = \infty$$

implies that $\sup_{t \in K} |Y(t)|$ is unbounded a.s.

II.) Let $0 \leq \beta < 1$. Then

(6.27) $$\int_0^\infty (\log \log N(k, d_{Y,\phi}(s,t); \varepsilon))^{1-\beta} d\varepsilon = \infty$$

implies that $\sup_{t \in K} |Y(t)|$ is unbounded a.s.

Proof: I.) follows from Theorem 1.2 with $\psi \sim \phi$ at infinity and $\nu(x) = x$ in (1.25). Note that in this case ϕ satisfies (1.44) with $p = 1$. II.) follows from Part I.) of Theorem 1.2, which remains valid as stated if the statement ϕ convex (see the paragraph containing (1.19)) is replaced by "ϕ is non-decreasing". ($d_{Y,\phi}$ is still defined as in (1.21) although it is no longer necessarily a pseudo-norm.) This is because the proof of Theorem 1.2 never uses the fact that ϕ is convex or that $d_{Y,\phi}$ is a metric. In applying part I.) of Theorem 1.2 in this case we take $\psi \sim \phi \sim \nu$.

Concerning sufficient conditions for continuity of the random Fourier series when $\psi(|\lambda|)$ is regularly varying near zero with index 1 let us note the for $\psi(\lambda) = \lambda$ we get a 1-stable random Fourier series which is a special case of a 1-stable ξ-radial process. The same difficulties mentioned above apply here.

We next consider consider ξ-radial processes with Levy transforms which are comparable to regularly varying functions of index 2 at infinity. As usual τ denotes the Levy measure and Ψ the Levy transform as defined in (2.2). Theorem 4, [BT] does not apply in this case. Instead we will use the following inequalities which are obtained the same way as the corresponding results for characteristic functions. Recall that $a \sim b$ means that there exist constants $0 < c_1$, $c_2 < \infty$ such that $c_1 a < b < c_2 a$.

$$(6.28) \qquad \tau\left[\tfrac{1}{u}, \infty\right) \leq \frac{7}{u} \int_0^u \Psi(\lambda) \, d\lambda \quad , \qquad\qquad \forall \, u > 0.$$

$$(6.29) \qquad \int_0^{1/3u} x\tau[x, \infty) \, dx < \frac{4}{u^3} \int_0^u \Psi(\lambda) \, d\lambda \quad , \quad \forall \, u > 0.$$

$$(6.30) \qquad \int_0^{\pi/u} x\tau[x, \infty) \, dx \geq \frac{\Psi(u)}{u^2} \quad , \qquad\qquad \forall \, u > 0.$$

We will give two different sufficient conditions for continuity. They complement Theorem 1.1 which does not contain conditions for

continuity for processes with regularly varying Levy transforms of index
2. Our first result, Theorem 6.5, gives an integral condition of the form
of (5.65). This Theorem requires far fewer smoothness conditions than
Theorem 6.6 which gives an integral condition of the form of (1.74) (or
equivalently (5.66)). Nevertheless, because it supports the conjecture in
Chapter I, Theorem 6.6 is of greater interest to us. Note that if
$\Psi(\lambda) \sim \lambda^2 (\log \lambda)^\delta$ as $\lambda \to \infty$ then δ must be less than or equal to 0
or else τ is not a Levy measure.

Theorem 6.5: Let $\{X(t)\}_{t \in K}$ be an R^K, $(\mathbb{C}^K)$ valued stochastic process with
characteristic functional given by (1.3) ((1.5)) and let Φ and $d_{X,\Phi}$ be
as defined in the paragraph containing (1.6). Suppose that $\Psi(|x|)$ is
regularly varying at infinity with index 2 such that there exists a
constant k $< \infty$ for which

(6.31) $\Psi(|x|) \leq k\Phi(x),$ $x \geq x_0'.$

Then if

(6.32) $\int_0^\infty \left(\log N(K, d_{X,\Phi}; \varepsilon)\right)^{1/2} d\varepsilon < \infty$

$\{X(t)\}_{t \in K}$ has a version with continuous sample paths.

Proof: For simplicity we will consider the real case. The complex case
is completely similar. By Lemma 2.3 we can represent $\{X(t)\}_{t \in K}$ by the
series

(6.33) $\sum_{j=1}^\infty \varepsilon_j F^{-1}(\Gamma_j) \gamma_j(t)$, $t \in K$

where $\{\gamma_j\}$ are i.i.d. Γ valued random variables distributed according to
the measure m. The other terms are defined prior to Lemma 2.3. We
already know that the series in (6.33) will converge uniformly a.s. iff
the series

(6.34)
$$\sum_{j=1}^{\infty} \varepsilon_j F^{-1}(j) \gamma_j(t), \ t \ \varepsilon \ K$$

converges uniformly a.s. Since the uniform convergence a.s. of (6.34)

depends on $F^{-1}(j)$ for j large it depends only on $\tau[x, \infty)$ for x near zero.

Therefore, as far as this Theorem is concerned, we can assume that the

Levy measure τ is supported on $[0, 1]$. (This does not interfere with the

assumption that Ψ is regularly varying at infinity.) In this case, by

(6.30) we see that $\overline{\lim_{u \to 0}} \ \Psi(|u|)/u^2 \leq C$ for some constant C. Thus we can

find a convex function $\tilde{\Phi}(x)$, $\tilde{\Phi}(0) = 0$ such that

(6.35)
$$\Psi(|x|) \leq k_1 \tilde{\Phi}(x) \ , \qquad \forall x \geq 0$$

and such that $\tilde{\Phi}(x) \leq k_2 \Phi(x)$, $x \geq x_0'' > 0$. This implies that

(6.36)
$$d_{X,\tilde{\Phi}}(s,t) \leq C d_{X,\Phi}(s,t) \ , \qquad \forall s,t \ \varepsilon \ K$$

for some constant C.

Let $\xi \geq 0$ be a random variable and define

(6.37)
$$\delta = \inf \left\{ t : \sum_{j=1}^{\infty} P\big(F^{-1}(j)\xi > t\big) < \frac{1}{72} \right\} .$$

Using (6.28) and (6.35) and the monotonicity of $\tilde{\Phi}$ we see that

(6.38)
$$\sum_{j=1}^{\infty} P\big(F^{-1}(j)\xi > t\big) \leq \sum_{j=1}^{\infty} P\big(\tau[t/\xi, \infty) > j\big)$$

$$\leq E \ \tau[t/\xi, \infty) \ \leq \ k \ E \ \tilde{\Phi}\Big(\frac{\xi}{t}\Big)$$

for some constant k independent of ξ. Define

(6.39) $\|\xi\|_{\tilde{\Phi}} = \inf \{c \cdot > 0: \ E \ \tilde{\Phi}(\frac{\xi}{c}) \leq 1\} \equiv \alpha$

and set $t = h\alpha$, $h \geq 1$ in (6.38). Then by the convexity of $\tilde{\Phi}$ we have by (6.38) and (6.39) that

$$\sum_{j=1}^{\infty} P(F^{-1}(j)\xi > t) \leq \frac{k}{h} E \ \tilde{\Phi}(\frac{\xi}{\alpha}) = \frac{k}{h} .$$

This shows that

(6.40) $\delta \leq h'\alpha$

for some h' sufficiently large which is independent of ξ.

It follows from the proof of Theorem 1.1, Chapter I, [MP1] that the series in (6.33) converges uniformly a.s. if

(6.41) $\int_0^{\infty} \left(\log N(K, \ Ed_{X,\tilde{\Phi};\omega} ; \ \varepsilon) \right)^{1/2} d\varepsilon < \infty$

where

$$d_{X,\tilde{\Phi};\omega}(s,t) = \left[\sum_{j=1}^{\infty} \left(F^{-1}(j)\right)^2 |\gamma_j(s) - \gamma_j(t)|^2 \right]^{1/2} .$$

Thus we can complete the proof of this Theorem by showing that

(6.42) $E \ d_{X,\tilde{\Phi};\omega}(s,t) \leq C'd_{X,\tilde{\Phi}}(s,t)$, $\forall s,t \in K$

for some constant C' and then using (6.36).

We will now obtain (6.42). Let $\{\xi_j\}_{j=1}^{\infty}$ be an i.i.d. sequence of random variables. By Khinchine's inequality and Theorem 3.3 [GZ] we have

(6.43) $E \left(\sum_{j=1}^{\infty} \left(F^{-1}(j)\right)^2 \xi_j^2 \right)^{1/2} \leq C \ E \left| \sum_{j=1}^{\infty} \varepsilon_j F^{-1}(j)\xi_j \right|$

$\leq C' \left[E \sup_j |F^{-1}(j)\xi_j| + \sum_{j=1}^{\infty} \left(F^{-1}(j)\right)^2 E\xi^2 \ I_{\left[F^{-1}(j)\xi < \delta\right]} \right]^{1/2}$

where δ is given in (6.37), $\{\varepsilon_j\}_{j=1}^{\infty}$ is a Rademader sequence and C and C' are constants. Note that

(6.44)
$$\sum_{j=1}^{\infty} \left(F^{-1}(j)\right)^2 E\xi^2 I_{\left[F^{-1}(j)\xi<\delta\right]}$$

$$= \sum_{j=1}^{\infty} \int_0^{\delta} uP\left(F^{-1}(j)\xi>u\right) du$$

$$\leq \int_0^{\delta} u \sum_{j=1}^{\infty} P(\tau[u/\xi, \infty) > j) du$$

$$\leq E \int_0^{\delta} u\tau[u/\xi, \infty) du$$

$$= E \xi^2 \int_0^{\delta/\xi} v\tau[v, \infty) dv .$$

Using (6.29), (6.40) and (6.35) and the convexity of $\tilde{\Phi}$, we see that this last term in (6.44)

(6.45)
$$\leq E \xi^2 \int_0^{h'\alpha/\xi} v\tau[v,\infty) dv$$

$$\leq 12k_1 h'\alpha^2 E \tilde{\Phi}\left(\frac{\xi}{\alpha}\right) = 12k_1 h'\alpha^2 ,$$

for some constant h' independent of ξ.

Finally we must evaluate

(6.46)
$$E \sup_j \left|F^{-1}(j) \xi_j\right|$$

$$\leq \int_0^{\infty} \left(1 \wedge \sum_{j=1}^{\infty} P(F^{-1}(j)\xi > u)\right) du$$

$$\leq \alpha + \int_{\alpha}^{\infty} \sum_{j=1}^{\infty} P(F^{-1}(j)\xi > u) du$$

$$\leq \alpha + E \int_{\alpha}^{\infty} \tau[u/\xi, \infty) du.$$

Since the support of τ is $[0,1]$ this last term

(6.47)
$$\leq \alpha + E \int_{\alpha\wedge\xi}^{\xi} \tau[u/\xi, \infty) du.$$

Now let $\tilde{\Psi}(\lambda) = \sup_{0 \leq x \leq \lambda} \Psi(x)$. It is well known, (see e.g. [Kl]) that $\tilde{\Psi}(\lambda)$ is also regularly varying at infinity of index 2. Using (6.28), (6.35), and the fact that $\tilde{\Psi}$ is of regular variation, we see that (6.47)

(6.48)
$$\leq \alpha + 7E \int_{\alpha \wedge \xi}^{\xi} \tilde{\Psi}\left(\frac{\xi}{u}\right) du$$

$$= \alpha + 7E \, \xi \left(\int_{1}^{\xi/\alpha \wedge \xi} \frac{\tilde{\Psi}(s)}{s^2} \right) ds$$

$$\leq \alpha + CE(\alpha \wedge \xi) \, \tilde{\Psi}\left(\frac{\xi}{\alpha \wedge \xi}\right)$$

$$\leq \alpha + Ck_1 \alpha E\left[\tilde{\Phi}\left(\frac{\xi}{\alpha}\right) + \tilde{\Phi}(1)\right]$$

$$\leq \alpha + C_1 \alpha \left(1 + \tilde{\Phi}(1)\right) \leq C_2 \alpha,$$

for constants C, C_1 and C_2 independent of ξ.

Consider (6.43) − (6.48). We see that

$$E \left(\sum_{j=1}^{\infty} \left(F^{-1}(j)\right)^2 \xi_j^2 \right)^{1/2} \leq C\|\xi\|_{\tilde{\Phi}}$$

for a constant C independent of ξ. Thus if we take ξ to the random variable $|Y(s) - Y(t)|$ distributed according to the measure m we have (6.42). This completes the proof of this Theorem.

Note that by (6.30), $\Psi(u) = o(u^2)$ as $u \to \infty$ thus $d_{X,\Phi}$ can always be taken smaller than L^2. For stationary Gaussian processes, as is well known, (6.32) is necessary and sufficient with $d_{X,\Phi}(s,t) = \left(E|Y(s) - Y(t)|^2\right)^{1/2}$.

Theorem 6.5 was not too difficult because the fact that ξ-radial processes are marginally subgaussian enabled us to use Theorem 1.1, [MP1]. Our next theorem is more complicated. We must follow the general approach developed in Chapter 4 for proving the sufficient conditions for continuity in Theorem 1.1. The reader will recall that when the Levy transform Ψ is regularly varying with index $1 < p < 2$ then

$\Psi(x) \sim \tau[\frac{1}{x},\infty)$ as $x \to \infty$ where τ is the Levy measure. This relationship

was used repeatedly in the proof of Theorem 1.1. When Ψ is regularly

varying with index 2 this relationship is no longer true. In fact if

$\tau[\frac{1}{x},\infty) \sim x^2(\log x)^{-\beta}$, $\beta > 1$ and $x \geq x_0$ sufficiently large then

$\Psi(x) \sim x^2(\log x)^{1-\beta}$ as $x \to \infty$. We nevertheless can relate Ψ and τ by

the general relationships (6.28)-(6.30) to which we add the following: If

$\tau[x,\infty)$ is convex for $0 < x \leq x'$ for some x' then

$$(6.49) \qquad \int_0^{\pi/(2u)} x\tau[x,\infty)dx \sim \frac{\Psi(u)}{u^2} \quad \text{as} \quad u \to \infty.$$

The next Theorem, with all its conditions, is still applicable in the

case of smooth Levy measures and transforms such as the ones mentioned

just above.

<u>Theorem 6.6</u>: Let $\{X(t)\}_{t\epsilon K}$ be an $\mathbb{R}^k(\mathbb{C}^k)$ valued stochastic process

with characteristic functional given by (1.3) ((1.5)). Let τ be the

Levy measure corresponding to Ψ and assume that $\tau[t,\infty)$ is regularly

varying at zero with index -2 and that there exist constants k, x_0 and

t_0 such that

$$(6.50) \qquad \tau[\frac{1}{xy},\infty) \leq k\tau[\frac{1}{x},\infty)y^2, \qquad \forall x \geq 1, \quad \forall y \geq 1$$

$$(6.51) \qquad \int_0^t u\tau[u,\infty)du \sim t^2\tau[t,\infty) \log 1/t, \quad 0 < t \leq t_0$$

and

$$(6.52) \qquad \tau[\frac{1}{x},\infty) \text{ is convex and } x^{-1}\tau[\frac{1}{x},\infty) \text{ is concave for } x \geq 1.$$

Assume furthermore that

$$(6.53) \qquad L(y \log y) \sim L(\frac{y}{\log y}) \qquad \text{as} \quad y \to \infty$$

where $L(x) = x^{1/2}F^{-1}(x)$ for F^{-1} as defined in Lemma 2.1. Then if

(6.54) $$J(H_\Psi, d_{X,2}) < \infty$$

$\{X(t)\}_{t \in K}$ has a version with continuous sample paths a.s. (Note that $\tau[\frac{1}{x}, \infty)$ convex for $x \geq 1$ implies that $\tau[t, \infty)$ is convex for $0 < t \leq 1$. Therefore by (6.49) and (6.51) we have that $\Psi(x) \sim \tau[\frac{1}{x}, \infty)\log x$ and so we can take $\Psi^{-1}(x) \sim [\tau^{-1}(x/\log x)]^{-1}$ in the definition of H_Ψ.)

Proof: There is one major deviation from the proof of Theorem 1.1, part II. We must obtain a variation of Lemma 4.1. Consider the material at the beginning of Chapter 4. Suppose that $\delta: \mathbb{N} \to \mathbb{R}$ defined there is regularly varying of index $1/2$ and satisfies

(6.55) $$\sum_{n=1}^{\infty} \frac{1}{\delta(n)^2} < \infty .$$

For $(\alpha_i)_{i \in I}$ complex numbers define $\|(\alpha_i)_{i \in I}\|_{\delta,\infty}$ as in (4.1). The remarks proceeding (4.1) show that $\|(\alpha_i)_{i \in I}\|_{\delta,\infty}$ is equivalent to a norm on $\ell_{\delta,\infty}(I)$. We need an analogue of the right hand inequality in Lemma 4.1 for the $(\delta(n))_{n \in \mathbb{N}}$ considered here. In this case the function q is no longer the one given in (4.2). In fact we find it too difficult to obtian a nice result like Lemma 4.1 and must introduce many smoothness conditions.

Lemma 6.7: Let $(\delta(n))_{n \in \mathbb{N}}$ be a regularly varying function of index $1/2$ satisfying (6.55) and such that

(6.56) $$\sum_{k=n}^{\infty} (\delta(k))^{-2} \sim n(\delta(n))^{-2} \log n \quad \text{as } n \to \infty.$$

Let $h(y)$ be a concave function satisfying

$$h(y) \geq \frac{y \log y}{\delta(y \log y)} \qquad y \geq y_0 > 1$$

for some constant y_0. Then for $(\alpha_i)_{i \in I} \; \varepsilon \; \ell_{\delta,\infty}(I)$

(6.57)
$$\| \sum_{i \in I} \varepsilon_i \alpha_i \|_{\psi_h} \leq C \| (\alpha_i)_{i \in I} \|_{\delta,\infty}$$

for some constant C where $\| \; \|_{\psi_h}$ is defined in (4.4) and (4.5).

Proof: We follow the proof of Lemma 4.1 and use (6.56) to obtain, for $n \geq 3$

(6.58) $\quad P(| \sum_{i \in I} \varepsilon_i \alpha_i | > 2C \frac{n}{\delta(n)}) \leq \exp - \frac{C^2 n^2}{2(\delta(n))^2} \; (\sum_{k > n} \delta(k)^{-2})^{-1}$

$$\leq \exp - C'n(\log n)^{-1}$$

for some constant C'. Setting $y = n(\log n)^{-1}$ we see that we can find constants C_1, C_1' and y_0 such that

(6.59)
$$P(| \sum_{i \in I} \varepsilon_i \alpha_i | > C_1 h(y)) \leq \exp - C_1' y .$$

The rest of the proof continues as in Lemma 4.1.

Continuation of the proof of Theorem: Let us first do the case where $\{X(t)\}_{t \in K}$ is real. Following the proof of Theorem 1.1, part II.) we recall that it is enough to prove the uniform convergence a.s. of

(6.60)
$$Y(t) = \sum_{n=1}^{\infty} \varepsilon_n F^{-1}(n) \gamma_n(t), \quad t \; \varepsilon \; K.$$

Since we are assuming (6.52) we have that τ and F^{-1} are inverses of each other. To adhere to the notation of Chapter 4 we set

$$\kappa(n) = \frac{1}{F^{-1}(n)} , \qquad n \; \varepsilon \; \mathbb{N}$$

Consider the marginal series in (4.75). We associate with this series the random metric

(6.61)
$$d_\omega(s,t) = \left\| \left(\frac{\gamma_n(s;\omega)-\gamma_n(t;\omega)}{\kappa(n)} \right)_{n\in N} \right\|_{\kappa(n),\infty}.$$

It follows by (6.57) and Lemma 4.2 with $\kappa(n) = \delta(n)$ that this marginal series is continuous a.s. (w.r.t. (Ω',F',P')), if

(6.62)
$$\int_0^{\hat{d}_\omega} h(\log N(K,d_\omega;\epsilon))d\epsilon < \infty$$

where $\hat{d}_\omega = \sup\limits_{s,t\in K} d_\omega(s,t)$ and h is a concave function satisfying

(6.63)
$$h(y) \geq (y \log y)F^{-1}(y \log y).$$

(Obviously we use Lemma 4.1 with $(\delta(n))^{-1} = F^{-1}(n)$. The fact that $\sum_n(F^{-1}(n))^2 < \infty$ is just (2.3)). Also note that by (6.52) the right side of (6.63) is actually concave for $y \log y \geq 1$ and so we can take $h(y) = (y \log y)F^{-1}(y \log y)$ for $y \log y \geq 1$.

The alternate expression for (6.61) is

(6.64)
$$d_\omega(s,t) = \sup_{n\geq 1} \kappa(n)\left(\frac{\gamma_n(s;\omega)-\gamma_n(t;\omega)}{\kappa(n)}\right)^*$$

from which we see that $\hat{d}_\omega \leq 2$. Continuing to follow the proof of Theorem 1.1 part II.) we see that the next major step is to compute $Ed_\omega(s,t)$. To do this we use Theorem 4.6. Extend $\kappa(n)$ so that $\kappa(x) = \dfrac{1}{F^{-1}(x)}$, $x \geq 1$. We see that

$$\kappa^{-1}(x) = \tau[\tfrac{1}{x},\infty).$$

If we choose $T(x) = \tau[\tfrac{1}{x},\infty)$ we see by (6.50) that (4.48) and (4.49) are satisfied with $\theta^{-1}(x) = \tau[\tfrac{1}{x},\infty)$ and $\Phi(y) = y^2$. Since $\theta(x) = \dfrac{1}{F^{-1}(x)} =$

$\kappa(x)$, (4.50) of Theorem 4.6 gives us

(6.65)
$$Ed_\omega(s,t) = E \sup_{n \geq 1} \kappa(n) \left(\frac{\gamma_n(s)-\gamma_n(t)}{\kappa(n)}\right)^*$$

$$\leq C\|\gamma(s) - \gamma(t)\|_2$$

for some constant C independent of the measure m (which defines γ). Using (6.65) in (4.80) we see that all we need to complete the proof of the Theorem (in the real case) is to show that

(6.66)
$$H_\psi(y) \geq y \log y \, F^{-1}(y \log y), \quad y \geq y_0$$

for some constant y_0. Recall that by definition

$$H_\psi(x) = \begin{cases} \int_1^x \dfrac{dx}{\psi^{-1}(x)} & x \geq 1 \\ \\ 0 & 0 \leq x < 1 \end{cases}$$

By the Note at the end of the statement of this Theorem and the fact that τ is regularly varying there exist constants u_0 and C such that

$$\psi^{-1}(u) \leq C(\tau^{-1}(u/\log u))^{-1}, \qquad u \geq u_0 .$$

Therefore there exists a constant k such that

$$H_\psi(y) \geq kyF^{-1}(y/\log y), \qquad y \geq y_0 .$$

Therefore, to obtain (6.66) we need to show that

$$(y \log y)F^{-1}(y \log y) \leq CyF^{-1}(y/\log y)$$

for some constant C. This is implied by (6.53) and completes the proof in the real case.

In the complex case there is nothing more to prove except that ψ and τ are restricted to be of type G, (see (2.27), (2.28) and the material immediately following.) This is not really a restriction in the

context of this Theorem since if τ satisfies the hypotheses of this

Theorem then $\tau_g = E_g \tau [\frac{x}{|g|}, \infty) \sim \tau [x, \infty)$, $x \geq 1$. This follows since

$E\tau_g [\frac{|x|}{g}, \infty) \geq \tau [x, \infty) P(|g| \geq 1)$ and by (6.50) $\tau_g [\frac{x}{|g|} I_{[|g| \geq 1]}, \infty) \leq$

$k\tau [x, \infty) |g|^2 I_{[|g| \geq 1]}$, $\forall x \geq 1$ whereas by the monotonicity of τ,

$\tau_g [\frac{|x|}{|g|} I_{[|g| \leq 1]}, \infty] \leq k\tau [x, \infty)$, $\forall x > 0$. Furthermore $\Psi_g(x) \sim \tau_g [\frac{1}{x}, \infty) \log x$

as above, so since $\tau_g \sim \tau$ we have $\Psi_g \sim \Psi$ by (6.30) and (6.49).

Furthermore it is not necessary that τ_g satisfy (6.50) and (6.52) for

all $x \geq 1$ but only that it is comparable at infinity to a function τ

that does. In other words one can consider Levy measures τ that satisfy

the hypotheses of this Theorem as being of type G.

Remark: Conditions (6.50) and (6.52) seem stringent in that these

relations are required to hold for all $x \geq 1$. A function like

$\tau [\frac{1}{x}, \infty) = x^2 (\log x)^{-\beta}$, $\beta > 1$, clearly will satisfy these relations for x

sufficiently large but not necessarily for all $x \geq 1$. This is really no

problem. Suppose that $\tau [\frac{1}{x}, \infty)$ satisfies (6.50) and (6.52) for $x \geq x_0$.

Then $\tau_0 = \tau [\frac{1}{x_0 x}, \infty)$ satisfies them for all $x \geq 1$. The Theorem can be

proved for τ_0. Then Lemma 7.3 and the fact that τ is regularly varying

show that the Theorem holds for τ. Let H_{Ψ_0} be the function analagous to

H_Ψ but obtained from τ_0. Then H_{Ψ_0} and H_Ψ are comparable because τ_0

and τ are.

Analagous to Corollary 5.4 we shall give some examples of ξ-radial

process with smooth Levy measures and probability measures m on the

group characters $\{e^{ikt}\}_{k=0}^\infty$. Both integral conditions (6.32) and (6.54)

will be shown to hold for the continuous processes and neither will hold

for the discontinuous processes. The situation here is slightly more

complicated since we are dealing with functions which are regularly

varying of index 2. Let $\tau [x, \infty)$ be a convex function $0 < x \leq 1$ with

$\tau [1, \infty) = 0$ and

(6.67) $\tau[x,\infty) = \left[x^2\left(\log \frac{1}{x}\right)^{1+\beta}\right]^{-1}$, $0 < x \le x_0 \le 1$

where $\beta > 0$. Since $\tau[x,\infty)$ is convex we have by (6.49) that

(6.68) $\Psi(u) \sim \dfrac{u^2}{(\log u)^{\beta}}$ as $u \to \infty$.

__Theorem 6.8__: Let $G = [0,2\pi]$ so that $\Gamma = \{e^{ikt}, k \in \mathbb{N}\}$. Let $\{X(t)\}_{t\epsilon G}$
be a $\mathbb{C}^G$ valued stochastic process with Levy measure given by (6.67) (and
consequently with Levy transform satisfying (6.68).) Let the measure m in
(1.5) satisfy

(6.69) $m\{e^{ikt}\} = C_\beta\left[k(\log k)^2(\log \log k)^{2-\beta+\epsilon}\right]^{-1} \equiv a_k$

for $k \ge 10$, $m\{e^{ikt}\} = C_\beta$, $0 \le k < 10$, where C_β is such that $|m| = 1$.
Then if $\epsilon > 0$ in (6.69), $\{X(t)\}_{t\epsilon G}$ has a version with continuous
sample paths. However, if $\epsilon \le 0$, $\sup_{t\epsilon G}|X(t)|$ is unbounded a.s.

__Proof__: We first obtain the condition for unboundedness. Unfortunately
(5.24) is not good enough in this case. We use a result of Giné and Zinn
which allows us to consider truncated variances.

Consider the function F^{-1} associated with τ in Lemma 2.3. For τ as
given in (6.67) we see that

(6.70) $F^{-1}(j) \sim \left[(j^{1/2}(\log j)^{\frac{\beta+1}{2}}\right]^{-1}$

for j sufficiently large. Also let

$$A_i = \{e^{ikt}: \ 2^i < k \le 2^{i+1}\}$$

and let $m_i = m(A_i)$. We see from (6.69) that

(6.71) $m_i \sim \left[i^2(\log i)^{2-\beta+\epsilon}\right]^{-1}$.

By (5.25) and (5.26) we see that $\sup\limits_{t \in G} |X(t)|$ is unbounded a.s. if

(6.72)
$$\sum_{i=1}^{\infty} E\left(\sum_{j=j_0}^{\infty} (F^{-1}(j))^2 \delta_{j,i}\right)^{1/2} = \infty$$

for some j_0 sufficiently large where $\{\delta_{j,i}\}_{j=1}^{\infty}$ is i.i.d. with $P(\delta_j = 1) = m_i$, $P(\delta_j = 0) = 1-m_i$.

It follows from Theorem 3.3, [GZ] that there exists a constant $C > 0$ such that

(6.73)
$$E\left|\sum_{j=j_0}^{\infty} \varepsilon_j F^{-1}(j) \delta_{j,i}\right|$$

$$\geq C\left(E\left|\sum_{j=j_0}^{\infty} \varepsilon_j F^{-1}(j) \delta_{j,i} I_{[F^{-1}(j) \delta_{j,i} \leq \gamma_i]}\right|^2\right)^{1/2}$$

where $\{\varepsilon_j\}_{j=1}^{\infty}$ is a Rademader sequence and

$$\gamma_i = \inf\left\{t: \sum_{j=j_0}^{\infty} P(F^{-1}(j)\delta_{j,i} > t) \leq \frac{1}{72}\right\}.$$

Let $t = F^{-1}(k)$, then

$$\sum_{j=j_0}^{\infty} P\left(\delta_{j,i} > \frac{F^{-1}(k)}{F^{-1}(j)}\right) = (k-j_0) m_i.$$

Therefore, for $i \geq i_0$ sufficiently large we can choose

(6.74)
$$\gamma_i \geq F^{-1}\left(\frac{c}{m_i}\right)$$

for some constant $c > 0$. Substituting (6.74) in (6.73) we have

(6.75)

$$E \left| \sum_{j=j_0}^{\infty} \varepsilon_j F^{-1}(j) \delta_{j,i} \right| \geq c \left(\sum_{j=j_0}^{\infty} (F^{-1}(j))^2 \ E\left(\delta_{j,i} I_{\left[F^{-1}(j)\delta_{j,i} \leq F^{-1}(\frac{c}{m_i})\right]} \right) \right)^{1/2}$$

$$\geq c \left(\sum_{j>\frac{c}{m_i}}^{\infty} (F^{-1}(j))^2 \ E\delta_{j,i} \right)^{1/2}$$

$$= (m_i)^{1/2} \left(\sum_{j>\frac{c}{m_i}}^{\infty} (F^{-1}(j))^2 \right)^{1/2} .$$

By the Schwarz Inequality there exists a constant $C_1 > 0$ such that

(6.76) $$E \left(\sum_{j=j_0}^{\infty} (F^{-1}(j))^2 \delta_{j,i} \right)^{1/2} \geq C_1 E \left| \sum_{j=j_0}^{\infty} \varepsilon_j F^{-1}(j) \ \delta_{j,i} \right|$$

Combining (6.76) and (6.75) we see that the series in (6.72) diverges whenever $\varepsilon \leq 0$ in (6.69). This verifies our assertion about the unboundness of $\{X(t)\}_{t \in G}$.

The continuity part of this Theorem is obtained by using Theorem 6.5. Note that, even though $\Psi(|x|)$ is required to be regularly varying at infinity in the statement of Theorem 6.5, we only used in the proof that $\Psi(|x|)$ is dominated by a regularly varying function for x sufficiently large which in turn satisfies (6.31). Thus, by (6.68) we can take

(6.77) $$\Psi(x) \leq k \ \frac{x^2}{(\log x)^\beta} \equiv \Phi(x) \quad , \quad \forall \ x \geq x_0.$$

As we discussed in the paragraph containing (5.73) and (5.74), the integral condition in (6.32) is equivalent to

(6.78) $$\int_0^\delta \frac{\overline{d_{X,\Phi}(u,0)}}{u(\log 1/u)^{1/2}} \ du < \infty$$

for $\delta > 0$ sufficiently small. Furthermore, by a change of variables (6.78) is equivalent to

(6.79)
$$\sum_{n=n_0}^{\infty} \frac{\overline{d_{X,\Phi}(1/n,0)}}{n(\log n)^{1/2}} < \infty .$$

Therefore to complete this portion of the proof we need only show that

(6.79a) $\overline{d_{X,\Phi}(1/n,0)} \leq c\big((\log n)^{1/2}(\log \log n)^{1+\epsilon/2}\big)^{-1} \equiv \alpha_n$

for those sequences $\{a_k\}$ in (6.69) which have $\epsilon > 0$. This is exactly
what is shown from (5.69) to (5.72) since that argument also works for
$p = q = 2$. In fact this Theorem is an exact extension of Theorem 1.5 to
the case $p = q = 2$, $\beta \geq 0$. (Note that in our notation we have replaced β
of Theorem 1.5 by $-\beta$.) Note also that in the argument of Chapter 5 a
specific value is chosen for all $x \geq 0$ whereas we have never specified
what our $\Phi(x)$ is for x small. Nevertheless, this doesn't matter because
the norms d_{X,Φ_1} and d_{X,Φ_2} are equivalent as long as $\Phi_1(x) \sim \Phi_2(x)$ as
$x \to \infty$. This completes the proof of Theorem 6.8.

Let us note that the integral in (6.32) is finite for those processes
in Theorem 6.8 which are continuous and infinite for those processes in
Theorem 6.8 which are unbounded. Indeed in the final part of the proof of
Theorem 6.8 we showed the first of these statements, that is if $\epsilon > 0$ in
(6.69) then the integral in (6.32) is finite. To see that if $\epsilon \leq 0$ in
(6.69) the integral in (6.32) is infinite we just follow the argument in
the paragraph containing (5.75) to (5.80) since that argument remains
valid for $p = q = 2$ and $\beta \leq 0$. (In the notation of this Chapter we use $-\beta$
in place of the term β used in Chapter 5.)

The integral in (6.54), which is the same as (1.74) with $p = 2$, is
also finite for the continuous processes in Theorem 6.8 and infinite for
those which are discontinuous. This is because (6.54) is equivalent to

$$\sum_{n=n_0}^{\infty} \frac{\overline{d_{X,2}(1/n,0)}}{n(\log n)^{1/2}(\log \log n)^{\beta/2}} < \infty$$

and

$$\overline{d_{X,2}(1/n,0)} \sim \left(\sum_{k=n}^{\infty} a_k \right)^{1/2} \sim \left[(\log n)^{1/2} (\log \log n)^{1-\beta/2+\epsilon/2} \right]^{-1}$$

as $n \to \infty$. These statements are proved employing the same arguments used repeatedly in Chapter 5. Thus the examples in Theorem 6.8 support the conjecture in Chapter 1 but again we have the same curious similarlity between (6.32) and (6.54) as between (5.65) and (5.66). (Note that $\tau[x,\infty)$ in (6.67) satisfies all the conditions of Theorem 6.6.

Theorem 1.1 also gives a necessary condition (of the form of (1.61) with $p = q = 2$) for the processes considered in Theorem 6.8 but it is easy to check that it is not best possible.

We now pass on to the random Fourier series and once again have a whole set of results which parallel those obtained for the ξ-radial processes. Let ξ and $\psi(\lambda)$ be as defined in (1.17). The following inequalities are the analogues of those given in (6.28) - (6.30):

$$(6.80) \qquad P\left(|\xi| > \frac{1}{u}\right) \leq \frac{14}{u} \int_0^u \psi(\lambda) d\lambda \quad , \qquad\qquad \forall\, u > 0 ,$$

$$(6.81) \qquad \int_0^{1/3u} x P\left(|\xi| > x\right) dx < \frac{8}{u^3} \int_0^u \psi(\lambda) d\lambda \quad , \qquad \forall\, u > 0 ,$$

$$(6.82) \qquad \int_0^{\pi/u} x P\left(|\xi| > x\right) dx \geq \frac{\psi(u)}{u^2} \quad , \qquad\qquad \forall\, u > 0 .$$

Let us also note that when $P(|\xi| > x)$ is convex

$$(6.83) \qquad \int_0^{1/u} x P\left(|\xi| > x\right) dx \sim \frac{\psi(u)}{u^2} \quad , \qquad\qquad \forall\, u > 0.$$

Note that if $\psi(|\lambda|) \sim \lambda^2 (\log 1/|\lambda|)^\delta$ as $\lambda \to 0$ then δ must be greater than or equal to zero or else $e^{-\psi(\lambda)}$ is not a characteristic function. The next Theorem gives a sufficient condition for continuity when $\psi(|\lambda|)$ is regularly varying at zero with index 2.

<u>Theorem 6.9</u>: Consider $\{Y(t)\}_{t\in K}$ as defined in (1.18) and ϕ and $d_{Y,\phi}$ as defined in the paragraph proceeding Theorem 1.2. Suppose that $\psi(|\lambda|)$ is regularly varying at zero with index $p = 2$ and that there exists a constant k' for which

(6.84) $\psi(|x|) \le k'\phi(x),\ 0 \le x \le x_0$

and

(6.85) $\overline{\lim_{x \to 0}}\ \frac{\psi(x)}{x^2} = \infty$.

Then if

(6.86) $\int_0^\infty \left(\log N(K,\ d_{Y,\phi};\ \varepsilon)\right)^{1/2} d\varepsilon < \infty$

the series representing $\{Y(t)\}_{t\in K}$ given in (1.18) converges uniformly a.s. and is a continuous version of $\{Y(t)\}_{t\in K}$. (Note that if the limit in (6.85) is finite this theorem follows from Theorem 1.1. Chapter I, $[MP1]$ with $d_{Y,\phi}(s,t)$ replaced by $\left(E|Y(t) - Y(s)|^2\right)^{1/2}$.)

<u>Proof</u>: Note that by a renormalization which doesn't alter the problem we can take $x_0 = 1$ and $\phi(1) = 1$ in (6.84). By (5.45) and the monotonicity of ϕ we have for $b_k \le u$

(6.87) $\sum_{k=1}^\infty P\left(|b_k\,\xi_k| > u\right) \le 14 \sum_{k=1}^\infty \frac{u}{b_k} \int_0^{b_k/u} \phi(v)\,dv$

$\le 14k' \sum_{k=1}^\infty \phi\left(\frac{b_k}{u}\right)$

where $\{\xi_k\}_{k=1}^\infty$ are i.i.d. copies of ξ given in (1.17). Let $u = h\|\{b_k\}\|_\phi$ for $h \ge 1$ where

(6.88) $\|\{b_k\}\|_\phi = \inf\left\{c > 0 : \sum_{k=1}^\infty \phi\left(\frac{|b_k|}{c}\right) \le 1\right\} \equiv \beta$.

By the convexity of ϕ we have that the last term in (6.87)

(6.89) $$\leq \frac{14k'}{h} \sum_{k=1}^{\infty} \phi\left(\frac{|b_k|}{\beta}\right) \leq \frac{14k'}{h} \quad .$$

Now let

(6.90) $$\delta = \inf \left\{ t : \sum_{k=1}^{\infty} P(|b_k \xi_k| > t) < \frac{1}{72} \right\}.$$

By (6.87) and (6.89) we see that for $h \geq 1$ sufficiently large, but independent of the sequence $\{b_k\}_{k=1}^{\infty}$,

(6.91) $$\delta \leq h\beta = h\|\{b_k\}\|_\phi \quad .$$

Now let us consider $\{Y(t)\}_{t \in K}$ as defined in (1.18). It follows from the proof of Theorem 1.1, Chapter I, [MPI] that this series converges uniformly a.s. if

(6.92) $$\int_0^\infty \left(\log N(K, E \, d_{Y,\phi,\omega} ; \epsilon) \right)^{1/2} d\epsilon < \infty$$

where

(6.93) $$d_{Y,\phi;\omega}(s,t) = \left(\sum_{\gamma \in A} |a_\gamma|^2 |\xi_\gamma|^2 |\gamma(s) - \gamma(t)|^2 \right)^{1/2} \quad .$$

Thus we can complete this proof by showing that

(6.94) $$E \, d_{Y,\phi;\omega}(s,t) \leq C \, d_{Y,\phi}(s,t)$$

for some constant C independent of s and t. By Khinchine's inequality and Theorem 3.3 [GZ] we have

(6.95) $$E \left(\sum_{k=1}^{\infty} |b_k|^2 |\xi_k|^2 \right)^{1/2} \leq C \, E \sum_{k=1}^{\infty} |b_k \xi_k|$$

$$\leq C' \left[E \sup_k |b_k \xi_k| + E \left(\sum_{k=1}^{\infty} |b_k|^2 |\xi_k|^2 I_{[|b_k \xi_k| \leq \delta]} \right)^{1/2} \right]$$

for δ as given in (6.90). By (6.80), (6.84), (6.91) and the monotonicity of ϕ we see that the last term in (6.95)

$$(6.96) \qquad \leq \left(\sum_{k=1}^{\infty} |b_k|^2 \int_0^{h\beta/b_k} uP[\xi > u] \, du \right)^{1/2}$$

$$\leq C h \beta \left(\sum_{k=1}^{\infty} \phi\left(\frac{b_k}{h\beta}\right) \right)^{1/2}$$

$$\leq C h^{1/2} \beta \left(\sum_{k=1}^{\infty} \phi\left(\frac{b_k}{\beta}\right) \right)^{1/2}$$

$$= C h^{1/2} \beta \ .$$

Also

$$(6.97) \qquad E \sup_k |b_k \xi_k| \leq \int_0^{\infty} [1 \wedge \sum_{k=1}^{\infty} P(|b_k\xi_k| > u)] du$$

$$\leq h\beta + \sum_{k=1}^{\infty} \int_{h\beta}^{\infty} P(|b_k\xi_k| > u) \, du \ .$$

As in the proof of Theorem 6.5 we can assume that $\psi(u)$ is non-decreasing for $0 \leq u \leq 1$ and still is regularly varying at zero with index 2 and satisfies (6.84). Therefore, by (6.80), (6.84) and the regular variation of ψ, the last term in (6.97)

$$\leq C \sum_{k=1}^{\infty} \int_{h\beta}^{\infty} \psi\left(\frac{|b_k|}{u}\right) \, du$$

$$\leq C' h\beta \sum_{k=1}^{\infty} \phi\left(\frac{|b_k|}{h\beta}\right)$$

$$\leq C' \beta \sum_{k=1}^{\infty} \phi\left(\frac{|b_k|}{\beta}\right) = C'\beta$$

for constants C and C' independent of $\{b_k\}$. Therefore, recalling (6.88) we see that

$$(6.98) \qquad E \left(\sum_{k=1}^{\infty} |b_k|^2 |\xi_k|^2 \right)^{1/2} \leq C''\|\{b_k\}\|_\phi$$

for some constant C' independent of $\{b_k\}$. By an appropriate change of variables (6.98) gives (6.94). This completes the proof of Theorem 6.9.

We next give some examples for which the sufficient condition in Theorem 6.9 is sharp. For ξ and ψ as given in (1.17) assume that $P(|\xi| > x)$ is a convex function on $[0, \infty)$ and

$$(6.99) \qquad P(|\xi| > x) = \frac{(\log x)^{\beta-1}}{x^2} \ , \ x \geq x_0 > 0$$

for $\beta > 0$. By (6.83) this implies that

$$(6.100) \qquad \psi(|u|) \sim u^2 (\log 1/u)^{\beta} \qquad \text{as } u \to 0 \ .$$

Theorem 6.10: Let $G = [0, 2\pi]$ so that $\Gamma = \{e^{ikt}, k \in \mathbb{N}\}$. Consider

$$(6.101) \qquad Y(t) = \sum_{k=0}^{\infty} a_k \xi_k e^{ikt}, \qquad t \in [0, 2\pi]$$

where $\{\xi_k\}_{k=0}^{\infty}$ are i.i.d. copies ξ defined in (6.99). (This implies (6.100)). Let

$$(6.102) \qquad a_k = \left[k^{1/2} (\log k)^{1+\beta/2} (\log \log k)^{1+\epsilon} \right]^{-1}$$

$k \geq 10$, otherwise let $a_k = 1$. Then if $\epsilon > 0$ in (6.102) the series in (6.101) converges uniformly a.s. whereas if $\epsilon \leq 0$, $\sup_{t \in G} |Y(t)|$ is unbounded a.s.

Proof: By Theorem 3.3 [GZ] there exists a constant $C > 0$, independent of $\{a_k\}$ and $\{\xi_k\}$ such that

$$(6.103) \qquad E \left| \sum_{k=2^n+1}^{2^{n+1}} a_k \xi_k \right| \geq C \left(E \sum_{k=2^n+1}^{2^{n+1}} |a_k \xi_k|^2 \ I_{\left[|a_k \xi_k| \leq \delta_n \right]} \right)^{1/2}$$

where

$$\delta_n = \inf \left\{ t : \sum_{k=2^n+1}^{2^{n+1}} P(|a_k \xi_k| > t) < \frac{1}{72} \right\} \ .$$

Using the given values of $\{a_k\}$ and ξ we see that

$$\delta_n \geq c_2 \left[n^{3/2} (\log n)^{1+\epsilon} \right]^{-1} , \qquad n \geq n_0$$

for some constant $c_2 > 0$ independent of n and n_0, for some n_0 sufficiently

large. Thus we see that the right side of (6.103)

(6.104)
$$\geq 2^n a_{2^{n+1}}^2 E|\xi|^2 I_{\left[a_{2^n} \xi \leq \delta_n \right]}$$

$$\geq c_3 \left(n^2 (\log n)^{2+2\epsilon} \right)^{-1} , \qquad n \geq n_0$$

for some constant $c_3 > 0$ independent of n and n_0. Since by Khinchine's

inequality

(6.105)
$$E \left(\sum_{k=2^{n+1}}^{2^{n+1}} |a_k \xi_k|^2 \right)^{1/2} \geq C' E \left| \sum_{k=2^{n+1}}^{2^{n+1}} a_k \xi_k \right|$$

we can use (6.105) together with (6.103), (6.104) and (5.11) to verify our

assertion on unboundedness.

We use Theorem 6.9 to verify the assertion on continuity. The proof

is similar to the proof of the continuity part of Theorem 6.8. In fact we

only need to obtain (6.79) with $d_{Y,\phi}$ replacing $d_{X,\phi}$. Clearly, we need

only show that $\overline{d_{Y,\phi}(1/n,0)} \leq \alpha_n$ for α_n given in (6.79a). This is done

in the paragraph containing (5.83) to (5.86) and the reader can readily

check that the argument given there remains valid for $p = 2$ and $\beta > 0$.

This completes the proof of Theorem 6.10.

Note that the integral in (6.86) is finite for those processes in

Theorem 6.10 which are continuous and infinite for those processes in

Theorem 6.10 which are unbounded. In fact we used Theorem 6.9 to prove

the continuity part of Theorem 6.10. To see that if $\epsilon \leq 0$ in (6.102) the

integral in (6.86) is infinite we just follow the final argument in

Chapter 5, in particular (5.106) and the paragraph containing (5.111), and

note that the argument works equally well when $p = 2$ and $\beta > 0$. On the

Michael B. Marcus

other hand it is easy to find unbounded lacunary random Fourier series for which (6.86) holds.

Theorem 1.2 also gives a necessary condition for continuity that can be applied to the processes considered in Theorem 6.10. However it is easy to see that it is also not best possible.

7. SUPREMA OF ξ-RADIAL PROCESSES AND RANDOM FOURIER SERIES

The series representation of ξ-radial processes enables us to estimate the asymptotic distribution of the sup-norms of these processes when they are continuous. These estimates are particularaly sharp when $\tau[x,\infty)$ is regularly varing at infinity where, as usual, τ denotes the Levy measure. We begin by developing some useful comparison results for these processes.

Lemma 7.1: Let

$$(7.1) \qquad X(t) = \sum_{j=1}^{\infty} \epsilon_j F^{-1}(\Gamma_j) Y_j(t), \qquad t \in T$$

be a real valued ξ-radial process as defined in (2.5). If the series in (7.1) converges uniformly a.s., (equivalently, if $(X(t))_{t \in T}$ has a version with continuous sample paths) then so does the series

$$(7.2) \qquad X_a(t) = \sum_{j=1}^{\infty} \epsilon_j F^{-1}(a\Gamma_j) Y_j(t), \qquad t \in T$$

(equivalently, then $(X_a(t))_{t \in T}$ has a version with continuous sample paths) for all $0 < a < \infty$.

Proof: By definition $F^{-1}(u)$ is non-increasing in u. Thus if $a \geq 1$ this lemma is an immediate consequence of the contraction principle. Let $\{X_i(t))_{t \in T}\}_{i=1}^{n}$ be independent copies of $(X(t))_{t \in T}$. If $(X(t))_{t \in T}$ has a version with continuous sample paths then so does $(X_1(t) + \dots + X_n(t))_{t \in T}$. (The equivalence between $(X(t))_{t \in T}$ having a version with continuous smaple paths and the series (7.1) converging uniformly a.s. is discussed in the proof of Lemma 2.7.) By checking the

164   Michael B. Marcus

characteristic functional one can see that $(X_1(t) + \ldots + X_n(t))_{t\epsilon T}$ is equivalent to the ξ-radial process

(7.3)
$$\sum_{j=1}^{\infty} \epsilon_j F^{-1}(\frac{\Gamma_j}{n}) Y_j(t), \quad t \epsilon T .$$

Furthermore by the above remarks this series has a version with continuous sample paths if $(X_1(t) + \ldots + X_n(t))_{t\epsilon T}$ does. This completes the proof of the lemma since one can interpolate using the contraction principle.

There are two components in the definition of a real valued ξ-radial process as defined in (2.5). A symmetric real valued infinitely divisible random variable ξ and a probability measure m on the unit ball of R^T. The next lemma shows that under certain conditions on ξ the ξ-radial process always has a version with continuous paths. This is a known result [AG] but we prove it here as an elementary consequence of the series representation.

Lemma 7.2: Let $(X(t))_{t\epsilon T}$ be a ξ-radial process as defined in (2.5) and let τ be the Levy measure of ξ as in (2.1) and (2.2). Then if

(7.4)
$$\int_0^{\infty} (1 \wedge t) d\tau [t,\infty) < \infty$$

$(X(t))_{t\epsilon T}$ has a version with continuous paths.

Proof: By the strong law of large numbers $\lim_{j\to\infty} \frac{\Gamma_j}{j} = 1$ a.s. Using this fact and Lemma 7.1 along with the representation of $(X(t))_{t\epsilon T}$ in (2.5) we see that $(X(t))_{t\epsilon T}$ has a version with continuous sample paths iff the series

$$\sum_{j=1}^{\infty} \epsilon_j F^{-1}(j) Y_j(t), \quad t \epsilon T$$

converges uniformly a.s. Clearly a sufficient condition for this is

$$\sum_{j=1}^{\infty} F^{-1}(j) < \infty$$

and this, by a change of variables, is equivalent to (7.4).

The next lemma is a contraction principle for ξ-radial processes, (see also [R]).

Lemma 7.3: Let $(X_1(t))_{t\epsilon T}$ and $(X_2(t))_{t\epsilon T}$ be two ξ-radial processes with the same spectral measure m. Let τ_1 and τ_2 be the corresponding Levy measures and suppose that

(7.5) $$\tau_1[\lambda,\infty) > c\tau_2[\lambda,\infty), \quad 0 < \lambda \leq \delta$$

for constants $c, \delta > 0$. Then if $(X_1(t))_{t\epsilon T}$ has a version with continuous sample paths so does $(X_2(t))_{t\epsilon T}$.

Proof: Assume first that $\delta = \infty$ in (7.5). Let $(X_2'(t))_{t\epsilon T}$ be the ξ-radial process with the same spectral measure m but with the Levy measure $c\tau_2[\lambda,\infty)$. If, following Lemma 2.3, we represent

(7.6) $$X_2(t) = \sum_{j=1}^{\infty} \epsilon_j F_2^{-1}(\Gamma_j)Y_j(t), \quad t \epsilon T$$

then we can represent

(7.7) $$X_2'(t) = \sum_{j=1}^{\infty} \epsilon_j F_2^{-1}(\Gamma_j/c)Y_j(t), \quad t \epsilon T$$

where $F_2^{-1}(\Gamma_j/c) = \sup[u: c\tau_2[u,\infty) > t]$. It follows from (7.5) that

$$F_1^{-1}(t) = \sup[u: \tau_1[u,\infty) > t] \geq F_2^{-1}(t/c).$$

Therefore by the contraction principle $(X_2'(t))_{t\epsilon T}$ has a version with continuous sample paths. But then, by Lemma 7.1, $(X_2(t))_{t\epsilon T}$ also has a

version with continuous sample paths. It is an elementary fact that the
continuity of a ξ-radial process only depends on the Levy measure near the
origin; thus it is only necessary to stipulate (7.5) for $0 < \lambda \le \delta$.
Further clarification of this last comment will be given immediately
below. This completes the proof of this lemma.

As is customary in the study of infinitely divisible processes it is
instructive to write a real valued ξ-radial process as a sum of two
independent ξ-radial processes. Let $(X(t))_{t \epsilon T}$ be a real valued ξ-radial
process as defined in (2.1) – (2.3) and (2.5). Such a process is defined
with respect to a symmetric real valued random variable ξ and a
probability measure m on the unit ball of R^T. Let $\tau[t,\infty)$, $0 < t < \infty$,
be the Levy measure corresponding to ξ and define

(7.8)
$$\tau_1[t,\infty) = \begin{cases} \tau[t,\infty) - \tau[1,\infty) & 0 < t < 1 \\ \\ 0 & t \ge 1 \end{cases}$$

and

(7.8a)
$$\tau_2[t,\infty) = \begin{cases} \tau[1,\infty) & 0 < t < 1 \\ \\ \tau[t,\infty) & t \ge 1 \end{cases}$$

Let Ψ_1 and Ψ_2 be the Levy transforms of τ_1 and τ_2 (as defined in
(2.2)) and ξ_1 and ξ_2 the symmetric infinitely divisible random
variables with Levy measures τ_1 and τ_2 (as defined in (2.1) and
(2.2)). Clearly $\xi \overset{D}{=} \xi_1 + \xi_2$ if ξ_1 and ξ_2 are independent. Let
$(X_i(t))_{t \epsilon T}$, $i = 1,2$ be the real-valued ξ-radial process with
characteristic functional given by

(7.9) $E \exp i \sum_{t \epsilon T} \alpha(t) X_i(t) = \exp - \int \Psi_i \left(\left| \sum_{t \epsilon T} \alpha(t)\beta(t) \right| \right) m(d\beta)$, $i = 1,2$

where the spectral measure m is the same as the spectral measure of
$(X(t))_{t \epsilon T}$. If $(X_i(t))_{t \epsilon T}$, $i = 1,2$ are independent then clearly

(7.10) $X(t) = X_1(t) + X_2(t), \quad t \in T$

in the sense of equivalence of stochastic processes. It is well known and

easy to see that if m is supported by C(T), the space of continuous

function on T, then $(X(t))_{t\in T}$ has a version with continuous sample

paths iff $(X_1(t))_{t\in T}$ does. This is because $(X_2(t))_{t\in T}$ is continuous.

A proof using the series representation goes as follows: Let

(7.11) $F_i^{-1}(t) = \sup[u: \tau_i[u,\infty) > t], \quad i = 1,2,.$

By Lemma 2.3

(7.12) $X_2(t) = \sum_{j=1}^{\infty} \varepsilon_j F_2^{-1}(\Gamma_j) Y_j(t), \quad t \in T$

where $F_2^{-1}(u)$ is zero for $u > \tau[1,\infty)$. Thus, almost surely, the series

(7.12) contains only a finite number of non-zero terms and we get a

continuous version of $(X_2(t))_{t\in T}$. (By exactly the same argument for m

on R^T, $(X(t))_{t\in T}$ has a bounded version iff $(X_1(t))_{t\in T}$ does.)

It is interesting to note that whereas $(X_1(t))_{t\in T}$ determines the

continuity or boundedness of $(X(t))_{t\in T}$ the distribution of the tail of

$\| X \|_{\infty} \equiv \sup_{t\in T} |X(t)|$, when it is finite, is generally the same as that of

$\| X_2 \|_{\infty}$. More explicitly we have the following:

Theorem 7.4: Let $(X(t))_{t\in T}$ be a ξ-radial process with spectral measure

m satisfying $\| X \|_{\infty} < \infty$ a.s. Let $(X_i(t))_{t\in T}$, ξ_i, τ_i, Ψ_i and F_i^{-1},

i = 1,2, be as defined above. In addition we define ξ_2^+ to be the

non-negative infinitely divisible random variable with characteristic

function

(7.13) $E\ e^{i\lambda \xi_2^+} = \exp - \int_1^{\infty} (e^{i\lambda t} - 1) d\tau_2[t,\infty).$

Suppose

(7.14) $\exists t_0 \in T$ such that $m[Y(t_0) = 1] = 1.$

Then, $\forall \delta < 1,$ $\lambda > 0$

(7.15) $P(|\xi_1| > \lambda) \leq P(\|X_1\|_\infty > \lambda) \leq P(N > \delta\lambda)$

where $N \geq 0$ is a random variable satisfying

(7.16) $E \exp cN(\log (e + N))^{1/2} < \infty$

and c is a constant independent of m. Also

(7.17) $P(|\xi_2| > \lambda) \leq P(\|X_2\|_\infty > \lambda) \leq P(\xi_2^+ > \lambda)$

and for all $0 \leq \varepsilon \leq 1$

(7.18) $P(|\xi| > \lambda) \leq P(\|X\|_\infty > \lambda) \leq P(\xi_2^+ > (1-\varepsilon)\lambda) + P(N > \varepsilon\delta\lambda).$

Without requiring (7.14) the right hand inequalities in (7.15), (7.17) and (7.18) remain valid and we always have

(7.19) $$\lim_{\lambda \to \infty} \frac{P(\|X\|_\infty > \lambda)}{\tau [\lambda,\infty)} \geq 1/2.$$

Proof: By (7.12), Lemma 2.2 and the fact that $\|Y_1\|_\infty = 1$ we have

$$\|X_2\|_\infty \leq \sum_{j=1}^\infty F_2^{-1}(\Gamma_j) \overset{\mathcal{D}}{=} \xi_2^+ .$$

This gives us the right side of (7.17). For the left side we simply note that by (7.12), (7.14) and Lemma 2.1

(7.20) $$\|X_2\|_\infty \geq \left| \sum_{j=1}^\infty \varepsilon_j F_2^{-1}(\Gamma_j) Y_j(t_0) \right|$$

$$= \left| \sum_{j=1}^\infty \varepsilon_j F_2^{-1}(\Gamma_j) \right| \overset{\mathcal{D}}{=} |\xi_2| .$$

Let us now consider $(X_1(t))_{t \in T}$. By Lemma 2.3 we can write

$$(7.21) \qquad X_1(t) = \sum_{j=1}^{\infty} \varepsilon_j F_1^{-1}(\Gamma_j) I_{[\Gamma_j \leq j/e^2]} Y_j(t)$$

$$+ \sum_{j=1}^{\infty} \varepsilon_j F_1^{-1}(\Gamma_j) I_{[\Gamma_j > j/e^2]} Y_j(t), \qquad t \in T.$$

Denote the sums in (7.21) in order by $(W_1(t))_{t \in T}$ and $(Z_1(t))_{t \in T}$. Since $\|Y_j\|_{\infty} \leq 1$ we have that

$$(7.22) \qquad \|W_1\|_{\infty} \leq \left[\sum_{j=1}^{\infty} \left(F_1^{-1}(\Gamma_j) \right)^2 \right]^{1/2} \left[\sum_{j=1}^{\infty} I_{[\Gamma_j \leq j/e^2]} \right]^{1/2}.$$

Let $\eta = \sum_{j=1}^{\infty} \left(F_1^{-1}(\Gamma_j) \right)^2$. We know that this sum converges a.s. by (2.3) and the strong law of large numbers. Furthermore, as we showed in the proof of Lemma 2.6, η is a non-negative, infinitely divisible random variable which has a Levy measure with compact support. It is well known, (see e.g. [K3]) that

$$(7.23) \qquad P(\eta > \lambda) \leq k e^{-k_1 \lambda L^+ \lambda}, \qquad \forall \lambda > 0$$

for constants k and k_1 where $L^+ \lambda \equiv \max(1, \log \lambda)$. Also, recognizing that $\Gamma_j = X_1 + \ldots + X_j$ is a random walk it follows from problem 12, Sec. 8.5 [C], that $\nu = \sum_{j=1}^{\infty} I_{[\Gamma_j < j/e^2]}$ satisfies

$$(7.24) \qquad P(\nu > \lambda) \leq k' e^{-k_1' \lambda}, \qquad \forall \lambda > 0$$

for constants k' and k_1'. (We also use the estimate

$$P[\Gamma_j < j/e^2] \leq \int_0^{j/e^2} \frac{x^{j-1}}{(j-1)!} dx = \frac{j}{j!} e^{-2j} < e^{-j}$$

which follows from (2.10).) Using (7.22) - (7.24) we have that

(7.25) $P(\|W_1\|_\infty > \lambda) \leq P\left(\eta > \dfrac{\lambda}{(L^+\lambda)^{1/2}}\right) + P(\nu > \lambda(L^+\lambda)^{1/2})$

$$\leq ce^{c_1\lambda(L^+\lambda)^{1/2}} ,$$

for constants c and c_1. Also by Theorem 3.2 [A]

(7.26) $P(\|Z_1\|_\infty > \lambda) \leq c'e^{-c_1'\lambda(L^+\lambda)} , \qquad \forall \lambda \geq 0$

for constants c' and c_1'. Combining (7.25) and (7.26) we get (7.16).

The right side of (7.18) follows immediately from the right sides of (7.15) and (7.17). The left side of (7.18) is obtained in the same manner as (7.20).

Finally to obtain (7.19) we note that since F_1^{-1} is non-increasing we have, using a contraction lemma for probabilities (see e.g. Lemma 1.3 (iii) [MZ]), that for all $\lambda > 0$,

(7.27) $P[\|\sum_j \varepsilon_j F^{-1}(\Gamma_j)Y_j(t)\|_\infty > \lambda] \geq \dfrac{1}{2} P[F^{-1}(\Gamma_1) > \lambda]$

$$= \dfrac{1}{2}\left(1 - e^{-\tau[\lambda,\infty)}\right),$$

since $P[\Gamma_1 \leq a] = 1 - e^{-a}$. The last term in (7.27) tends towards $\dfrac{1}{2}\tau[\lambda,\infty)$ as $\lambda \to \infty$.

Under some smoothness conditions we can specify the asymptotic distribution of $\|X\|_\infty$.

Corollary 7.5: Under the hypotheses of Theorem 7.4 including (7.14) suppose that $\tau[\lambda,\infty)$ is regularly varying at infinity with index $p < 0$. Then

(7.28)
$$\lim_{\lambda \to \infty} \frac{P(\| X \|_\infty > \lambda)}{\tau [\lambda, \infty)} = 1.$$

<u>Proof</u>: Let ξ_2' and $\xi_2^{+'}$ be independent copies of ξ_2 and ξ_2^+. Using characteristic functions it is easy to check that

(7.29)
$$\xi_2 + \xi_2' \overset{D}{=} \xi_2^+ - \xi_2^{+'}.$$

Also, it is well known, see e.g. Proposition 0, [EG] that

(7.30)
$$\lim_{\lambda \to \infty} \frac{P[\xi_2^+ > \lambda]}{\tau [\lambda, \infty)} = 1$$

where, obviously, it doesn't matter whether we use τ or τ_2 in (7.30).

Note that (7.30) along with (7.18) and (7.19) already give us

(7.31)
$$1/2 \le \lim_{\lambda \to \infty} \frac{P(\| X \|_\infty > \lambda)}{\tau [\lambda, \infty)} \le 1$$

without requiring (7.14). The remainder of the proof is to get the correct limit in (7.28).

It follows from Example (c), VIII.8 [F] and (7.29) that

(7.32)
$$\lim_{\lambda \to \infty} \frac{P(| \xi_2^+ - \xi_2^{+'} | > \lambda)}{\tau [\lambda, \infty)} = \lim_{\lambda \to \infty} \frac{P(| \xi_2 + \xi_2' | > \lambda)}{\tau [\lambda, \infty)} = 2.$$

Furthermore, by (8.17) of this same Example (c), VIII.8 [F] we have for all $\epsilon > 0$, $\lambda > 0$

(7.33)
$$P(\xi_2 + \xi_2' > \lambda) \le 2P(\xi_2 > \lambda(1-\epsilon)) + P^2(\xi_2 > \lambda \epsilon)$$
$$\le 2P(\xi_2 > \lambda(1-\epsilon)) + P^2(\xi_2^+ > \lambda \epsilon)$$

where we used (7.17) for the final inequality. Following the argument in [F] we see that

$$\lim_{\lambda \to \infty} \frac{P(\xi_2 > \lambda)}{\tau[\lambda,\infty)} = 1/2.$$

Therefore, since ξ_2 is symmetric,

(7.34) $$\lim_{\lambda \to \infty} \frac{P(|\xi_2| > \lambda)}{\tau[\lambda,\infty)} = 1.$$

Using (7.34) and (7.30) in (7.17) we get

$$\lim_{\lambda \to \infty} \frac{P(\|X_2\|_\infty > \lambda)}{\tau[\lambda,\infty)} = 1.$$

This, along with (7.15) gives (7.28).

In this proof critical use is made of Proposition 0, [EG] (see 7.30) in which it is required that $\tau[\lambda,\infty)$ is regularly varying at infinity. Results of this type but with weaker conditions on τ can be found in [HK] and can be used to obtain (7.28) for a wider class of Levy measures.

Remark 7.6: The division of τ into τ_1 and τ_2 at $\lambda = 1$ is completely arbitrary. Any $\lambda = a > 0$ would give the same results. Theorem 7.4 and Corollary 7.5 show that the problem of continuity or boundedness of ξ-radial processes and the distribution of the sup-norm of continuous or bounded ξ-radial are completely separate whenever the tail of the distribution of the sup-norm goes to zero more slowly than that of the random variable N, (see (7.15) and (7.16)). The question of what is the best estimate for $\|X_1\|_\infty$ is very interesting. The results in [A] suggest that (7.16) should hold with 1/2 replaced by 1, i.e. that $\|X_1\|_\infty$ has the same tail behavior as some multiple of a Poisson random variable, however we can not prove this. (Added in proof: A. de Acosta has informed me that he can obtain this result from a more general one that he has recently obtained by modifying the proof in [A]. He also said that M. Talagrand told him that he has solved this problem in the still more

general setting of infinitely divisible measures on a Banach space in
which the Levy measure is supported on a ball of finite radius.)

We now consider the distribution of the sup-norm of random Fourier
series as defined in (1.18) when the series converge uniformly a.s. By
Hoffmann-Jorgensen's inequality [HJ] (see also [JMI]) one has

$$(7.35) \qquad \varlimsup_{t \to \infty} \frac{P\left(\| \sum_{\gamma \in A} a_\gamma \xi_\gamma \gamma \|_\infty > 3t \right)}{P\left(\sup_{\gamma \in A} |a_\gamma \xi_\gamma| > t \right)} \leq 1 \; .$$

Also, by Levy's inequality

$$(7.36) \qquad \varliminf_{t \to \infty} \frac{P\left(\| \sum_{\gamma \in A} a_\gamma \xi_\gamma \gamma \|_\infty > t \right)}{P\left(\sup_{\gamma \in A} |a_\gamma \xi_\gamma| > t \right)} \geq 1/2.$$

Unfortunately the denominator in (7.35) or (7.36) is not easy to calculate
in general. We will briefly discuss this in the case when ψ is
regularly varying at zero with index $0 < p < 2$. Since, in this case,

$$(7.37) \qquad \lim_{t \to \infty} \frac{P(|\xi_\gamma| > t)}{\psi(1/t)} = C_p$$

for C_p a constant depending only on p, we might ask when is
$P(\sup_{\gamma \in A} |a_\gamma \xi_\gamma| > t) \sim \psi(1/t)$ as $t \to \infty$? We will give some conditions that
imply this behavior and some examples in which $P(\sup_{\gamma \in A} |a_\gamma \xi_\gamma| > t)$ is not
comparable to $\psi(1/t)$ as $t \to \infty$.

Lemma 7.6: Let ξ and ψ be as in (1.17) and assume that ψ is
regularly varying at zero with index $0 < p < 2$. Let $\{\xi_k\}_{k=1}^\infty$ be
i.i.d. copies of ξ and $\{a_k\}_{k=1}^\infty$ be real or complex numbers. Then

$$(7.38) \qquad \varliminf_{t \to \infty} \frac{P(\sup_k |a_k \xi_k| > t)}{\psi(1/t)} \geq C_p \sum_{k=1}^\infty |a_k|^p$$

where C_p is given in (7.37). In addition suppose that

(7.39) $$\psi(xy) \leq c\psi(x)\psi(y), \quad 0 < x \leq x_0; \quad 0 < y \leq y_0$$

for some $x_0, y_0 > 0$ and that $\{a_k\}_{k=1}^{\infty}$ is such that

(7.40) $$\sum_{k=1}^{\infty} \psi(|a_k|) < \infty.$$

Then

(7.41) $$\lim_{t \to \infty} \frac{P\left(\sup_k |a_k \xi_k| > t\right)}{\psi(1/t)} = C_p \sum_{k=1}^{\infty} |a_k|^p < \infty.$$

Proof: Since ψ is regularly varying at zero

(7.42) $$\lim_{t \to \infty} \frac{\psi(|a_k|/t)}{\psi(1/t)} = |a_k|^p .$$

Thus by (7.37) and (7.42) we have

(7.43) $$\lim_{t \to \infty} \frac{P(|a_k \xi_k| > t)}{\psi(1/t)} = C_p |a_k|^p.$$

It is well known that when $\lim_{t \to \infty} \sum_{k=1}^{n} P(|a_k \xi_k| > t) = 0$

(7.44) $$\lim_{t \to \infty} \frac{\sum_{k=1}^{n} P(|a_k \xi_k| > t)}{P\left(\sup_{1 \leq k \leq n} |a_k \xi_k| > t\right)} = 1.$$

(Here n can also be ∞). Therefore by (7.43) and (7.44) we have

$$\lim_{t \to \infty} \frac{P\left(\sup_{1 \leq k \leq n} |a_k \xi_k| > t\right)}{\psi(1/t)} = C_p \sum_{k=1}^{n} |a_k|^p$$

which gives us (7.38).

When (7.39) holds we have

$$\psi(|a_k|/t) \leq c\psi(|a_k|)\psi(1/t) \quad .$$

Therefore, using (7.40) we see that $\lim\limits_{t\to\infty} \sum\limits_{k=1}^{\infty} \psi(|a_k|/t) = 0$, which by (7.37) gives us that $\lim\limits_{t\to\infty} \sum\limits_{k=1}^{\infty} P(|a_k\xi_k| > t) = 0$ and this shows that under (7.39) we have (7.44).

To obtain (7.41) we note that by (7.37) and (7.44)

$$(7.45) \qquad \overline{\lim_{t\to\infty}} \frac{P\left(\sup\limits_{k} |a_k\xi_k| > t\right)}{\psi(1/t)} = \overline{\lim_{t\to\infty}} \sum_{k=1}^{\infty} \frac{P(|a_k\xi_k| > t)}{\psi(1/t)}$$

$$= C_p \overline{\lim_{t\to\infty}} \sum_{k=1}^{\infty} \frac{\psi(a_k/t)}{\psi(1/t)} \quad .$$

By (7.39)

$$\frac{\psi(a_k/t)}{\psi(1/t)} \leq c\psi(a_k).$$

Therefore by (7.40), (7.42) and the dominated convergence theorem applied to the last term in (7.45) we get

$$(7.46) \qquad \overline{\lim_{t\to\infty}} \frac{P\left(\sup\limits_{k} |a_k\xi_k| > t\right)}{\psi(1/t)} \leq C_p \sum_{k=1}^{\infty} |a_k|^p$$

which together with (7.38) gives us the equality in (7.41). We complete the proof by showing that (7.39) implies that $\sum\limits_{k=1}^{\infty} |a_k|^p < \infty$.

Since ψ is regularly varying of index p it can be represented in the form

$$(7.47) \qquad\qquad \psi(x) = x^p L(x)$$

where

$$(7.48) \qquad\qquad \lim_{y\to 0} \frac{L(xy)}{L(y)} = 1 \qquad \forall x > 0$$

By (7.39) we have

(7.49) $$\frac{L(xy)}{L(y)} \leq cL(x)$$

and taking the limit on the left as $y \to 0$ and using (7.48) shows us that

$L(x) \geq \frac{1}{c}$. Thus by (7.47), $\psi(x) \geq \frac{1}{c} x^p$ which implies, because of

(7.40), that $\sum_{k=1}^{\infty} |a_k|^p < \infty$. This completes the proof of Lemma 7.6.

In the next lemma we explore this problem when ψ satisfies a different growth condition at the origin.

Lemma 7.7: Let ξ and ψ be as in (1.17) and assume that ψ is regularly varying at zero with index $0 < p < 2$. Let $\{\xi_k\}_{k=1}^{\infty}$ be i.i.d. copies of ξ and $\{a_k\}_{k=1}^{\infty}$ real or complex numbers. Furthermore assume that

(7.50) $\psi(xy) \leq Cx^p\psi(y)$, $0 < x \leq x_0$, $0 < y \leq y_0$

for some $x_0, y_0 > 0$, and that $\{a_k\}_{k=1}^{\infty}$ is such that

(7.51) $$\sum_{k=1}^{\infty} \psi(|a_k|) < \infty .$$

Then

(7.52) $$\varlimsup_{t \to \infty} \frac{P\left(\sup_k |a_k\xi_k| > t\right)}{\psi(1/t)} \leq C \sum_{k=1}^{\infty} |a_k|^p$$

and

(7.53) $$\varlimsup_{t \to \infty} t^p P\left(\sup_k |a_k\xi_k| > t\right) \leq C \sum_{k=1}^{\infty} \psi(|a_k|) .$$

Proof: As in the proof of Lemma 7.6; (7.50), (7.51) and (7.37) show that
$\lim_{t \to \infty} \sum_{k=1}^{\infty} P\left(|a_k\xi_k| > t\right) = 0$. Therefore (7.44) holds. The rest is trivial since (7.50) implies both

(7.54) $$\psi\left(\frac{|a_k|}{t}\right) \leq c|a_k|^p \psi(1/t)$$

and

(7.55) $$\psi\left(\frac{|a_k|}{t}\right) \leq \frac{c\psi(|a_k|)}{t^p}$$

which gives us (7.52) and (7.53) respectively.

Note that (7.50) implies that $\psi(|a_k|) \leq c|a_k|^p$. The right side of (7.52) can be infinite when the right side of (7.53) is finite.

Remark 7.8: (7.53) gives an upper bound for the behavior of $P\left(\sup_k |a_k\xi_k| > t\right)$ as $t \to \infty$ that is probably best possible, but it is not very sharp in general. We will show this by giving examples. We choose sequences $\{a_k\}_{k=1}^{\infty}$ and $\{\xi_k\}_{k=1}^{\infty}$ for which the associated random Fourier series defined in (1.18) converges uniformly. Let

$$(7.56) \qquad a_k = \left[k^{1/p}(\log k)^{1 + \frac{\beta+\alpha}{p}}\right]^{-1}, \quad -\infty < \beta < \infty, \quad \alpha > 0, \quad 1 < p < 2$$

where

$$(7.57) \qquad\qquad p + \beta + \alpha \leq 1$$

and let

$$P\left(|\xi_k| > t\right) = \frac{(\log t)^\beta}{t^p}, \quad t \geq e .$$

By (7.37) and Corollary 5.5 we see that $\{Y(t), t \in [0, 2\pi]\}$, which is defined in (5.63), converges uniformly a.s. Also let us note that $\sum_{k=1}^{\infty} |a_k|^p = \infty$. We have for $j \geq j_0$ sufficiently large

$$\sum_{k=1}^{\infty} P\left(|a_k\xi_k| > j\right) = \sum_{k=1}^{\infty} \frac{|a_k|^p}{j^p} \left(\log \frac{j}{a_k}\right)^\beta$$

$$\sim \left| \frac{(\log j)^\beta}{j^p} \sum_{k=1}^{j} \frac{1}{k(\log k)^{p+\beta+\alpha}} + \sum_{k=j}^{\infty} \frac{1}{k(\log k)^{p+\alpha}} \right|$$

$$\sim \frac{1}{j^p(\log j)^{p+\alpha-1}} .$$

Therefore, by (7.44) we have

$$(7.58) \qquad\qquad P\left(\sup_k |a_k\xi_k| > t\right) \sim \frac{1}{t^p(\log t)^{p+\alpha-1}}$$

as $t \to \infty$, which doesn't even depend on β. By (7.37)

$$\psi(1/t) \sim P(|\xi_k| > t) \sim \frac{(\log t)^\beta}{t^p} \ .$$

Thus, neither the left side of (7.52) nor the left side of (7.53) gives the correct limiting behavior in these examples. This same phenomenon is also evident when $p = 1$, however we need to define the $\{a_k\}_{k=1}^\infty$ based on (6.17).

Remark 7.9: When $P(|\xi| > t)$ is regularly varying with index -2 it is not comparable to $\psi(1/t)$. To write Lemmas 7.6 and 7.7 for this case let $h(t) = P(|\xi| > t)$. Lemmas 7.6 and 7.7 are valid with $\psi(1/t)$ replaced by $h(t)$ and for $h(t)$ regularly varying with index $-\infty < p < 0$. This indicates that $h(t)$ is really the natural function to put in the denominator in (7.38), (7.41) and (7.52). We used $\psi(1/t)$ because Corollary 5.5 is given in terms of ψ.

REFERENCES

[A]   de Acosta, A., Strong exponential integrability of sums of
        independent B-valued random vectors, Probability and Math.
        Statistics, 1, (1980), 133-150.

[AG]   Araujo, A. and E. Giné, The central limit theorem for real and
        Banach valued random variables, (1980), Wiley, New York.

[B]   Belyaev, Yu. K., Continuity and Hölder's conditions for sample
        functions of stationary Gaussian processes, Proc. Fourth Berkeley
        Symp. Math. Statist. Prob. 2 (1961), 22-33.

[B1]   Breiman, L., Probability (1968), Addison-Wesley, Reading, Mass.

[BT]   Bingham, N. H. and J. L. Teugels, Tauberian theorems and regular
        variation, Nieuw. Arch. Wisk. (3) 27 (1979), no. 2, 153-186.

[C]   Chung, K. L., A course in probability theory (1968), Harcourt Brace,
        New York.

[CL]   Cuzick, J. and T. L. Lai, On random Fourier series, Trans. Amer.
        Math. Soc., 261 (1980), 58-80.

[D]   Dudley, R. M., The sizes of compact subsets of Hilbert space and
        continuity of Gaussian processes, J. Funct. Anal. 1 (1967), 290-330.

[EG]   Embrechts, P. and Goldie, C. M., Comparing the tail of an infinitely
        divisible distribution with integrals of its Lévy measure, Ann. of
        Probability, 9 (1981), 468-481.

[F]   Feller, W., An introduction to probability theory and its
        applications, Vol. II. First edition (1966), J. Wiley & Sons, New
        York.

[F1]   Fernique, X., Régularité des trajectoires des fonctions aléatoire
        gaussiennes, Lecture Notes in Mathematics, 480 (1975), 1-96,
        Springer-Verlag, New York.

[F2]   Fernique, X., Regularité des fonctions aléatoires non gaussiennes.
        Lecture Notes in Math. 976 (1983), 1-74, Springer Verlag, N.Y.

[GZ]   Giné, E. and Zinn, J., Central limit theorems and weak laws of large
        numbers in certain Banach spaces, Z. Wahrscheinlichkeitsth., 62,
        (1983), 323-354.

[HJ]   Hoffmann-Jørgensen, J., Sums of independent Banach space valued
        random variables, Studia Math., 52 (1974), 159-186.

[HK]   Hahn, M. and M. Klass, Uniform bounds for log-tail probabilities of
        infinitely divisible laws, Ann. of Probability, Special Invited
        Paper, in preparation.

[IN]   Ito, K. and M. Nisio, On the convergence of sums of independent
        Banach space valued random variables, Osaka Math. J. 5 (1968),
        35-48.

[JM]   Jain, N. C. and Marcus, M. B., Continuity of sub-gaussian processes,
        Advances in Probability, Vol. 4, (1978), M. Dekker, New York.

[JM1]   _____, Integrability of infinite sums of
        vector-valued random variables, Trans. Amer. Math. Soc., 212 (1975),
        1-36.

[K]   Kahane, J. P., Some random series of functions (1968), D. C. Heath,
        Lexington, Mass., Second Edition (1985), Cambridge University Press.

[K1]   Karamata, J., Sur un mode de croisance reguliere, Mathematica
        (Cluj), 4, (1930), 38-53.

[K2]   Kuelbs, J., Some exponential moments of sums of independent random
        variables, Trans. Amer. Math. Soc., 240 (1978), 145-162.

[K3]   Kruglov, V. M., A note on infinitely divisible distributions,
        Theor. Probability Appl. 15 (1970), 319-324.

[KR]   Krasnosel'skii, M. A. and Rutickii, Ya. B., Convex functions and
        Orlicz spaces, (1961), P. Noordhoff LTD., Groningen, Netherlands.

[L]   Loeve, M., Probability Theory (1955), D. Van Nostrand, Princeton,
        New Jersey.

[LM]   Ledoux, M. and Marcus, M. B., Some remarks on the uniform
        convergence of gaussian and Rademacher Fourier quadratic forms,
        Lecture Notes in Mathematics, 1193 (1986), 53-72.

[LeP]   Le Page, R., Multidimensional infinitely divisible variables and
        processes, Part II, Lecture Notes in Mathematics, 860 (1981),
        279-284.

[M]   Marcus, M. B., Continuity and the central limit theorem for random
        trigonometric series, Z. Wahrscheinlichkeitsth. 42 (1978), 35-56.

[MP1]   Marcus, M. B. and G. Pisier, Random Fourier series with applications
        to harmonic analysis, Ann. Math. Studies 101 (1981), Princeton
        Univ. Press, Princeton, N.J.

[MP2]   _____, Characterisations of almost surely continuous p-stable
        random Fourier series and strongly stationary processes, Acta Math.,
        Vol. 152 (1984), 245-301.

[MP3]   _____, Some results on the continuity of stable processes and
        the domain of attraction of continuous stable processes, Ann. Inst.
        Henri Poincaré, 20, (1984), 177-199.

[MZ]   Marcus, M. B. and Zinn, J., The bounded law of the iterated
        logarithm for the weighted empirical distribution process in the
        non-i.i.d. case, Ann. of Probability, 12 (1984), 335-360.

[NN]   Nanopoulis, C. and Nobelis, P., Etude de la regularité des fonctions
        aléatoire et de leur proprietés limites. These de 3e cycle
        (1977), Strasbourg.

[P]   Pisier, G., Some applications of the metric entropy conditions to
        harmonic analysis, in Banach Spaces, Harmonic Analysis and
        Probability Proceedings 80-81. Lecture Notes in Math. 995 (1983),
        123-154, Springer-Verlag, New York.

[P1]   Pitmann, E. J. G., Some theorems on characteristic functions of
        probability distributions, Proc. Fourth Berkeley Symposium, II,
        Univ. of California Press, Berkeley, (1961), 393-402.

[PZ]   Paley, R.E.A.C. and Zygmund, A., On some series of functions (1),
        (2), (3), Proceedings of Cambridge Phil. Soc., 26 (1930), 337-357,
        26 (1930), 458-474, 28 (1932), 190-205.

[R]   Rosinski, J., Random Integrals of Banach space valued functions,
        Studia Math, 78 (1984), 15-38.

[SZ]   Salem, R. and A. Zygmund, Some properties of trigonometric series
        whose terms have random signs, Acta Math. 91 (1954), 245-301.

This research was carried out while Professor Marcus was at Texas A&M
University, College Station, Texas. His current address is The City
College of CUNY, New York, NY 10031.

General instructions to authors for
PREPARING REPRODUCTION COPY FOR MEMOIRS

> For more detailed instructions send for AMS booklet, "A Guide for Authors of Memoirs."
> Write to Editorial Offices, American Mathematical Society, P. O. Box 6248,
> Providence, R. I. 02940.

MEMOIRS are printed by photo-offset from camera copy fully prepared by the author. This means that, except for a reduction in size of 20 to 30%, the finished book will look exactly like the copy submitted. Thus the author will want to use a good quality typewriter with a new, medium-inked black ribbon, and submit clean copy on the appropriate model paper.

Model Paper, provided at no cost by the AMS, is paper marked with blue lines that confine the copy to the appropriate size. Author should specify, when ordering, whether typewriter to be used has PICA-size (10 characters to the inch) or ELITE-size type (12 characters to the inch).

Line Spacing – For best appearance, and economy, a typewriter equipped with a half-space ratchet – 12 notches to the inch – should be used. (This may be purchased and attached at small cost.) Three notches make the desired spacing, which is equivalent to 1-1/2 ordinary single spaces. Where copy has a great many subscripts and superscripts, however, double spacing should be used.

Special Characters may be filled in carefully freehand, using dense black ink, or INSTANT ("rub-on") LETTERING may be used. AMS has a sheet of several hundred most-used symbols and letters which may be purchased for $5.

Diagrams may be drawn in black ink either directly on the model sheet, or on a separate sheet and pasted with rubber cement into spaces left for them in the text. Ballpoint pen is *not* acceptable.

Page Headings (Running Heads) should be centered, in CAPITAL LETTERS (preferably), at the top of the page – just above the blue line and touching it.

LEFT-hand, EVEN-numbered pages should be headed with the AUTHOR'S NAME;

RIGHT-hand, ODD-numbered pages should be headed with the TITLE of the paper (in shortened form if necessary).

Exceptions: PAGE 1 and any other page that carries a display title require NO RUNNING HEADS.

Page Numbers should be at the top of the page, on the same line with the running heads.

LEFT-hand, EVEN numbers – flush with left margin;

RIGHT-hand, ODD numbers – flush with right margin.

Exceptions: PAGE 1 and any other page that carries a display title should have page number, centered below the text, on blue line provided.

FRONT MATTER PAGES should be numbered with Roman numerals (lower case), positioned below text in same manner as described above.

MEMOIRS FORMAT

> It is suggested that the material be arranged in pages as indicated below.
> Note: <u>Starred items</u> (*) <u>are requirements of publication.</u>

Front Matter (first pages in book, preceding main body of text).

Page i – *Title, *Author's name.

Page iii – Table of contents.

Page iv – *Abstract (at least 1 sentence and at most 300 words).

*1980 Mathematics Subject Classification (1985 Revision). This classification represents the primary and secondary subjects of the paper, and the scheme can be found in Annual Subject Indexes of MATHEMATICAL REVIEWS beginning in 1984.

Key words and phrases, if desired. (A list which covers the content of the paper adequately enough to be useful for an information retrieval system.)

Page v, etc. – Preface, introduction, or any other matter not belonging in body of text.

Page 1 – Chapter Title (dropped 1 inch from top line, and centered).

Beginning of Text.

Footnotes: *Received by the editor date.

Support information – grants, credits, etc.

Last Page (at bottom) – Author's affiliation.